Y0-BQT-454

Bibliography of the History of Electronics

by

GEORGE SHIERS

assisted by

MAY SHIERS

The Scarecrow Press, Inc.
Metuchen, N.J. 1972

Library of Congress Cataloging in Publication Data

Shiers, George.
Bibliography of the history of electronics.

1. Electronics--History--Bibliography. I. Title.
Z5836.S54 016.621381'09 72-3740
ISBN 0-8108-0499-9

PREFACE

This bibliography contains 1820 listings of articles, books and other printed materials associated with the historical aspects of electronics and telecommunications. These include subject histories and companion works on science and physics, electricity, technology, inventions and industries that cover the period from around 1860 to recent years.

The first two chapters provide a liberal number of references to general works, literature guides and serials, all related to the major topics. Chapters 3 through 5 also contain works of a general nature on subject histories, biographies, and group histories. The remaining chapters are devoted to specific subjects. Details of the contents, topics and arrangement are given in the Introduction.

Various articles, pamphlets, books and other materials that were provided or made available to me by professional and commercial groups have proved useful in this work. It is a pleasure to record the assistance of the following companies and organizations: Armstrong Memorial Research Foundation, Inc., Bell Telephone Laboratories, A. C. Cossor Ltd., EMI Electronics Ltd., General Electric Company (Research and Development Center), General Radio Company, Institution of Electronic and Radio Engineers, International Telephone and Telegraph Corporation, Marconi Company, Palo Alto Chamber of Commerce, Radio Corporation of America, Radio Society of Great Britain, Raytheon Company, Television Society, Thorn-AEI Radio Valves and Tubes Ltd., U.S. Army Signal Corps (Fort Monmouth, N.J.), and Westinghouse Electric Company. For their help in furnishing these materials, I want to thank Colonel Donald L. Adams, Mr. R. W. Bell, Mr. Arnold M. Durham, Admiral F. R. Furth, Mr. Frank Griffin, Mr. F. C. G. Hiller, Mr. Kenyon Kilbon, Mr. Aaron Levine, Mr. R. V. McGahey, Mr. E. Brian Munt, Mr. Bruce Strasser, Mr. Thomas J. Styles, Mr. Peter Van Avery and Mr. Frederick Van Veen.

The facilities of the Library of the University of California, Santa Barbara, particularly the science and engineering division, have been essential in pursuing this work. I am most grateful to the staff for their services, especially Mr. Donald Fitch, Mr. John Johnson, Mrs. Lily Milton, and Mrs. Maria Patermann. For searching and finding many of the rarer books and less accessible journals, indexes and references, and for other important help, I am indebted to the staff of the Santa Barbara Public Library. Mrs. Thelma D. Evans, Mrs. Nadine Greenup, and Mrs. Corlice Wendling deserve special thanks for their efficient and friendly services and for their interest in my needs.

Personal thanks are extended to Mr. Gerald R. M. Garratt, Science Museum, London, Professor Charles Süsskind, University of California, Berkeley, and Mr. E. James Leaman, for their interest and support in various ways; to Mr. Vance Phillips for the loan of books, pamphlets, and catalogs related to early wireless telegraphy; to Mr. Paul E. Thompson for extensive loans of radio magazines not easily available; and to Mr. Elliot N. Sivowitch, Smithsonian Institution, who has frequently furnished important materials and much useful information.

I am grateful most of all to my wife, May, for her continued help in many ways; as a research assistant, typist, secretary, and reader, and for lightening the burden of sundry, and often tedious, tasks that have had to be done during the years while the manuscript was taking shape.

GEORGE SHIERS

Santa Barbara, California

CONTENTS

INTRODUCTION

The entries in this book cover rather more than one hundred years from the late 1850s. Chief subjects of concern are the histories of physics and electrical science, development of electron tubes, cathode-ray tubes and semiconductors; histories of technology and engineering with special reference to electronics, related electronic applications in electroacoustics, industry, computers and radar; and the progress of the main fields of telecommunications--telegraphy, telephony, radio and television.

Power engineering is excluded except for those histories of electrical engineering that also deal with telecommunications. Certain subjects, such as medical electronics, scientific applications and specialist fields that do not fit the general electronics and telecommunications categories, are also omitted.

The general reference works listed in the first chapter and the guides to serial publications in the second chapter were selected solely for their usefulness in the study of the history of science, physics, engineering, technology and electronics. Chapter 2 also contains a list of 91 magazines and journals related to the subject matter. There are 155 entries in Chapter 3 concerning general technical history. Biographical materials are listed in Chapter 4. In Chapter 5 there are 100 entries on the history of societies, institutions and commercial organizations. The remainder of the book consists of 11 chapters on the main subjects, arranged alphabetically from Broadcasting to Television.

The alphabetical sequence, which is the most useful for serial titles, real names and groups, is used in certain sections: 2E, Magazines and Journals; 4C, Individual biographies; and 5B through 5D, Group histories. In all other sections the entries are grouped in chronological order. This sequence has been maintained according to the publication date, or the end of the period covered in the entry, or the earliest date, whichever appeared to be most logical or convenient to the reader in relation to the topics covered in the particular entry and in the companion items.

Materials that treat multiple topics and those that survey a whole subject field or cover a lengthy period or deal generally with a particular era are grouped chronologically as introductions in Chapters 6, 8, 10, 11, 13, 16. Chapters 9, 10, 12, 13, 14 contain period subdivisions with the same chronological order. This arrangement, which complements the subject breakdown by sections, permits ready selection of entries for both topic and period.

The listing contains a variety of aids for reference, selection and study: bibliographies, catalogs, chronologies, indexes and literature guides. Other kinds of material range from condensations or brief reviews, such as surveys, addresses and synopses, to specialist works: memoirs, monographs, notable textbooks, professional papers and technical articles. Popular articles and books intended for the general reader, young adult or hobbyist, and a few business surveys are also listed to round out the historical approach or supplement the more technical records.

Several hundred items are absent because of their marginal historical value. Others were left out where it seemed they were merely repetitious or less inclusive than preferred items. A limited number of leading papers on original research, especially those that have come to be regarded as scientific or technical milestones, are included. There is also a limited number of textbooks that were considered major works of the times listed in most sections. However, these papers and textbooks were selected to be indicative of contemporary literature and technical progress and do not represent a comprehensive listing.

Although the approach is international, the contents are limited to works in English. Translations and English versions of foreign works are given wherever possible. Important contributions of workers outside England and the United States are recorded in some professional articles and books, usually documented with references to original papers and foreign patents. These are noted in a number of entries.

Essential bibliographic details are given in full for most entries. All these have been studied. Others which could not be obtained have been checked for reasonable accuracy and listed, even though incomplete, for the sake of representation. Citations include page numbers, page sequences, full date and volume number for articles; edition dates and page numbers for many books, as well as the usual place name and publisher. Special features, such as tables, maps, portraits, biographical lists, chronologies and appendixes are frequently noted, often with descriptions. Particulars of illustrations (usually overlooked) are emphasized; number, types, and sources, since these details are often quite useful in historical research. Similarly, the number and range of references, the earliest reference date, patent listings and other documentation details are noted where they may be of value.

Annotations are descriptive of the contents; particularly topics, period, treatment, coverage, level and limitations. Some entries are expanded with detailed comments that suit the technical importance or historical value of the works and indicate their range of interests and possible usefulness. The lack of an index or of references or other documentation is noted in a number of entries. These and other comments are given solely to guide the reader and are not intended as critical remarks on either the literary or technical quality of a work. In some instances, however, there are statements to indicate the technical stature or historical value, as for leading papers, basic books, or comprehensive works.

Notices of other relevant sections and numerous specific items listed elsewhere are given in the introductory text with most chapters and some sections. Cross references to related items have been supplied with numerous entries. Chapters and parts from nearly 100 items have been listed as subentries. Each one is identified with reference to the main entry and is included in the author index.

Some readers will be interested in electrical history before 1860 and in topics not covered in this bibliography. The senior work is Joseph Priestley's The History and Present State of Electricity, London, 1767. The third edition (1775) was recently reprinted (Johnson, New York, 1966, 2 vols.). Park Benjamin's The Intellectual Rise in Electricity (Wiley, New York; Longmans, Green, London, 1895) surveys events from early times to Franklin's kite experiments of 1752. Mottelay's Bibliographical History (236), with very extensive references, is the chief source for world literature covering all periods from B. C. 2637 and events to 1821; it also has references to publications after 1860.

Early telegraph history is treated fully in John J. Fahie's History of Electric Telegraphy to the Year 1837 (Spon, London, 1883).

The progress of electricity was a favorite topic for authors during the middle and latter part of last century. Some general periodicals, like Century, Leslies, Living Age, and Scribner's, featured articles, sometimes illustrated, on the history, progress and present state of electricity and its applications. Similar matter, usually more technical, appeared in the American Journal of Science (99), English Mechanic (131), Journal of the Franklin Institute (141), Nature (150), Popular Science Monthly (156), Scientific American (177), Scientific American Supplement (178), and in the older professional journals listed herein.

The following works are mentioned as prime sources for readers and students not acquainted with the literature. Section 1B contains several rich collections of books and other printed matter covering all phases and periods of electrical history: Ronald's Catalogue (7), Peddie's Subject Index (8), the Wheeler Catalog (12), and the Engineering Societies Subject Catalog (15). Publications of the Royal Society (79-82) list original works from 1800. Sources for twentieth-century material include Science Abstracts (83, 85), Engineering Index (84), Applied Science and Technology Index (88), and the "Critical Bibliography," Isis (29). Other bibliographies on specific topics and periods are listed or noted in the subject entries.

ABBREVIATIONS

Abbreviations used for organizations, companies and publications are not included. Periodical titles and abbreviations are listed in Sect. 2E.

App.	Appendix
Chap.	Chapter
c.	circa (about)
comp.	compiler
diags.	diagrams
ed.	editor, edition
engrs.	engravings
G. P. O.	Government Printing Office
H. M. S. O.	Her (His) Majesty's Stationery Office
illus.	illustrated, illustrations
No.	Number
NS	New Series
pp.	pages
Pt.	Part
port.	portrait
refs.	references
rev.	revised
Sect.	Section
Ser.	Series
Supp.	Supplement
trans.	translator, translated
Vol.	Volume

1. GENERAL REFERENCE WORKS

The entries in this chapter provide a broad base for general research, primarily in books and guides to the literature. Standard works covering general literature, such as the Britannica, Americana, other encyclopedias and reader's guides and handbooks have, with a few exceptions, been excluded These should not be overlooked, however, since they do contain articles, references and entries of potential value. Several entries in later chapters have been extracted from the Encyclopaedia Britannica, 11th ed., 1911. The year books issued as supplements to several encyclopedias and various annual compilations are other sources for dates, names, events, statistics and similar contemporary records. Although there is some overlapping of contents, works relating to periodicals are listed in Chapter 2.

A. General Bibliographies and Guides

1 Winchell, Constance M., Guide to Reference Books. Chicago: American Library Association, 1967. 8th ed., 741 pp. A comprehensive list of basic reference books published to 1964, classified by subject. Author-subject index, pp. 611-741. There are sections on bibliographies, catalogs, periodicals, government publications, biographies, and the pure and applied sciences (pp. 525-610), including physics, engineering and general scientific and technical works. Most entries are well annotated, with supplementary information on publication dates, editions, changes and so forth. First edition, 1902; subsequent editions 1908, 1917, 1923, 1929, 1936, 1951, with supplements for intervening years. Also current supplements with indexes.

2 The Bibliographic Index. A cumulative bibliography of bibliographies. New York: H. W. Wilson. First quarterly number March 1938; first annual volume 1939, 344 pp. Subject catalog with topical breakdown includes "History" entries also books and personal names as subjects. Typical periodicals indexed include Proceedings of the American Academy of Arts and Sciences, Annals of the American

Academy of Political and Social Science, American Journal of Science, Bell System Technical Journal, Electrical Engineering, General Electric Review, Journal of the Franklin Institute, Journal of the Institution of Electrical Engineers, Proceedings of the Institution of Radio Engineers, Proceedings of the National Academy of Sciences, Philosophical Magazine, Review of Scientific Instruments, Review of Modern Physics, Scientific Monthly, Smithsonian Institution Annual Report, Journal of the Western Society of Engineers, Wireless Engineer. Recent issue (1969), 477 pp.

3 Besterman, Theodore, A World Bibliography of Bibliographies and of Bibliographical Catalogues, Calendars, Abstracts, Digests, Indexes, and the Like. London: 1939-40, 2 vols. 2nd ed., 1947-49, 3 vols. 3rd ed., Geneva: Societas Bibliographica, 1955-56, 4 vols. 4th ed., Lausanne: Societas Bibliographica, 1965-66, 5 vols., including author and subject index volume. Two columns per page, 6664 columns, index 6683-8425. Arranged alphabetically by subject. Main subjects grouped in numbered sections, with initial index and column references. Entries in chronological sequence. The number of items in each bibliography is indicated. Pertinent listings (by volume and columns) include the following. Electricity and magnetism: 1:1925-1945. Electron, electronics, 1:1946-1947. Engineering, 2:1957-1971. Inventions, patents, 2:3089-3163. Radio, 4:5312-5317. Science, 4:5608-5672. Technology, 4:6014-6049. Telegraphy and telephony, 4:6054-6056. Television, 4:6056-6057. The complete listing of separate issues for certain entries, such as abstracts, patents, digests and similar progressive works, is a valuable feature. These detailed entries give the volume number, date or other coverage, number of pages and number of items.

4 Larson, Henrietta M., Guide to Business History. Materials for the Study of American Business History and Suggestions for Their Use. Cambridge, Mass.: Harvard University Press, 1948. Reprinted, Boston: J. S. Canner, 1964. 1181 pp. 4904 entries (some multiple); books, articles, government and other documents and reports, most with informative annotations. Topical classification in 87 groups. Index, pp. 1037-1181, by Elsie H. Bishop. There are only about 60 entries directly concerned with communications, electrical manufacturing, individual companies and the industry. Nevertheless, this work contains innumerable items of likely use, particularly about people and organizations, patents, government publications, and research and reference materials indirectly related to electronics history.

5 Hawkins, Reginald R. (ed.), Scientific, Medical, and Technical Books published in the United States of America: A Selected List of Titles in Print with Annotations. Washington: National Academy of Sciences, National Research Council, 1958. 2nd ed., 1491 pp. 7954 entries of books published to Dec. 1956. Full details, with list of contents, useful notes and comments. Directory of publishers, separate author and subject indexes.

6 Downs, Robert B., and Francis B. Jenkins (eds.), Bibliography, Current State and Future Trends. Chicago: University of Illinois Press, 1967. 611 pp. Although intended for the professional reference librarian, this compilation of 37 essays (originally published in Library Trends, Vol. 15, Jan., April, 1967) is a general survey and guide to a wide range of literature. Parts of particular interest treat periodicals, history of science, general science, physics, and engineering. There are 220 references, including foreign-language titles with these five chapters; three are listed elsewhere (see 25, 26, 39). The index refers to subjects, titles and authors.

B. Book Catalogs and Indexes

7 Ronalds, Francis (comp.), Catalogue of Books and Papers Relating to Electricity, Magnetism, The Electric Telegraph, etc., Including the Ronalds Library. London: E. & F. N. Spon, 1880. 564 pp. Biographical memoir of Sir Francis Ronalds, F.R.S., by Alfred J. Frost, pp. vii-xxvii. Over 13,000 entries of books, papers, pamphlets, monographs and miscellaneous material collected by Ronalds or noted by him from related works, booksellers' catalogs and similar sources. Entries are arranged alphabetically by author, with place and year of publication, number of pages and illustrations. Reference to the source is given with most of the extracted titles. Biographical data given for many authors. There is great emphasis on foreign-language works. This library was donated to the Society of Telegraph Engineers, predecessor of the Institution of Electrical Engineers, London.

8 Peddie, Robert A., Subject Index of Books Published up to and Including 1880. London: H. Pordes (reprint) 1962. 4 vols., each A-Z; 1933, 2nd series 1935, 3rd series 1939, new series 1948. 3419 pp. Brief entries of author, title, and date. A general index with subject cross references. There are sections on electrical engineering, electricity, patents, physics, and the telegraph.

9 Hiltebrand, E. (comp.), Subject Catalogue of the Memorial Library of the International Electrical Exhibition, held under the auspices of the Franklin Institute, September-October, 1884. Part of the Official Catalogue. Philadelphia: Burk and McFetridge, 1884. 97 pp. About 1500 entries, including cross-referenced entries. This collection of international titles includes unbound items, offprints, monographs, articles, periodicals and textbooks, chiefly donated by authors, publishers, societies and individuals. Many are of historical interest.

10 Fortescue, George K. (ed.), Subject Index of Modern Works added to the Library of the British Museum in the Years 1881-1900. London: 1902-1903. Three vols., A-E, F-M, N-Z, 2970 pp. Pertinent listings include electricity, pp. 799-815; telegraphy and the telephone, pp. 684-690. Continued in 5-year cumulations from 1901.

11 [U.K. Patent Office Library]. Electricity, Magnetism, and Electro-technics. London, 1904. Library Series No. 14, Subject list. 286 pp., about 2400 entries.

12 Weaver, William D., Catalogue of the Wheeler Gift of Books, Pamphlets and Periodicals in the Library of the American Institute of Electrical Engineers. With Introduction, Descriptive and Critical Notes by Brother Potamian. New York: The Institute, 1909, 2 vols. Vol. 1, 504 pp. Deed of gift, pp. v-vii. Introduction, pp. 17-46. Main portion of catalogue, pp. 49-504, 2447 entries from 1473 to 1906. Vol. 2, 475 pp. Excerpts from periodicals and miscellaneous, pp. 9-224 (2448-4378). Telegraph items, pp. 227-269 (4379-4828). Electric light, telephone and manufacturing companies, pp. 273-283 (4829-4985). Patents and litigation, pp. 287-302 (4986-5134). Legislation, pp. 305-326 (5135-5347). Expositions, etc., pp. 329-344 (5348-5507). Trade catalogues, etc., pp. 347-373 (5508-5852). Periodicals, pp. 377-405 (5853-5966). The sympathetic telegraph (with refs.), pp. 409-418. Author index, pp. 421-449. Index to telegraph entries, pp. 453-463. Report of library committee (1903), pp. 467-475. This extensive collection, built up over a period of 40 years by J. Latimer Clark, English electrical engineer (1822-1898), was presented to the A.I.E.E. by Dr. Schuyler Skaats Wheeler. Lists about 7000 items up to the end of the 19th century. Contains 114 plates from notable works in the collection.

13 [British Science Guild]. A Catalogue of British Scientific and Technical Books. Covering every Branch of Science and Technology carefully Classified and Indexed. London, 1921.

376 pp. Over 6000 entries in 50 main groups. Electrical engineering, pp. 109-121, includes 65 items on telegraphy, telephony and radio. 3rd ed., 1930, 754 pp., nearly 14,000 items. Succeeded by British Scientific and Technical Books. London: Aslib (by James Clarke), 1956. A select list of recommended books published in Great Britain and the Commonwealth in the years 1935 to 1952. 364 pp. About 8500 entries. Electrical engineering, pp. 123-141. Author or title index, pp. 271-343; subject index, pp. 344-352. List of British publishers, pp. 353-364. Same, 1953-1957, published 1960, 251 pp. About 11,000 entries. No annotations.

14 [New York State Library]. Checklist of Books and Pamphlets in Science and Technology. Albany, N.Y.: The Library, 1960. 216 pp. 81,800 items listed alphabetically by author with title, ed., year, D.D.C. call number. This finding list of the library's holdings includes communication, electronics, engineering, physics. Does not include periodicals or government documents.

15 [Engineering Societies Library]. Classed Subject Catalog. (New York.) Boston: G.K. Hall, 1963. 13 vols., including index vol. Includes all fields of engineering and related physical sciences, classified according to U.D.C. 10,631 pp., 242,000 cards, 21 card entries per page. Alphabetical index for subjects only, 356 pp., 26,700 entries. Yearly supplement volumes from 1964, each with subject index. There are brief notes with many cards. A major collection of 12 societies.

16 [Northern University, Boston Library]. A Selective Bibliography in Science and Engineering. Boston: G. K. Hall, 1964. 550 pp. Brief information on roughly 15,000 titles. Histories of science and technology, pp. 1-6.

17 [John Crerar Library]. Author-Title Catalog. (Chicago.) Boston: G.K. Hall, 1967. 35 vols., 28,554 pp., nearly 600,000 cards. Also, Classified Subject Catalog. 42 vols., 33,167 pp., 730,000 cards. Includes subject index in alphabetical order. This major collection covers all fields; engineering, physics, technology, including historical items.

C. Guides to Scientific and Technical Literature

18 Darrow, Karl K., Classified List of Published Bibliog-

raphies in Physics, 1910-1922. Washington: National Research Council, 1924. 102 pp.

19 Parke, Nathan G., Guide to the Literature of Mathematics and Physics: Including Related Works on Engineering Science. New York: McGraw-Hill, 1947, 205 pp. New York: Dover, 1958; London: Constable, 1959, 2nd rev. ed., 436 pp. Pt. 1, General considerations (74 pp.), on the principles of reading and study, self-directed education, literature search, periodicals. Pt. 2, The literature, over 5000 entries grouped alphabetically by subject. There is an introduction to each main subject group and also to numerous subgroups. Electronic and electrical engineering, pp. 194-209, contains 360 entries under 30 subgroups. Electronic physics, pp. 209-211, 65 entries under 7 subgroups. Foreign-language items included, but English titles not given. Author and subject indexes. Most items are from the 1920s to the early 1950s.

20 Whitford, Robert H., Physics Literature: A Reference Manual. Washington: Scarecrow Press, 1954, 228 pp. 2nd ed. (Metuchen, New Jersey); 1968, 272 pp. A thorough survey at the college level. Entries are grouped by approach; bibliographical, experimental, mathematical, educational, topical, etc. Much background material is furnished, with numerous extracts, comments and many footnotes. Chapter 3, Historical approach, pp. 45-59, has some 80 entries grouped under chronologies, bibliographies, physics histories, science histories, special histories, and current events. There are 10 items on electricity and magnetism. Other entries for electricity and magnetism are given in pp. 201-207; electronic items are listed in pp. 208-219. Foreign titles are given in the language of origin. There are separate author and subject indexes.

21 Fleming, Thomas P., Guide to the Literature of Science. New York: Columbia University, School of Library Service, 1957. 2nd ed., 69 pp. About 1000 items: books, periodicals, bibliographies, abstract sources, indexes. Engineering excluded.

22 Schutze, Gertrude, Bibliography of Guides to the S-T-M Literature: Scientific, Technical, Medical. New York: The Author, 1958. 64 pp., 654 entries. A comprehensive list of guides, general and topical, from 1920. Also supplements, 1963, 1967.

23 Moore, Charles K., and Kenneth J. Spencer, Electronics: A Bibliographical Guide. London: Macdonald; New York: Macmillan, 1961. 411 pp. More than 3000 entries classified by subject, with number of references and brief annotations, from 1945 to 1959. General refs., pp. 1-39, contains 300 items, including 28 history books and articles. Several dozen other items of historical interest are scattered throughout and noted in the index. Vol. 2, (London: Macdonald; New York: Plenum Press, 1965), 369 pp., covers the literature from 1959 to 1964. Historical entries, pp. 18, 19, 13 items, and elsewhere under subjects.

24 Jenkins, Frances B., Science Reference Sources. Champaign, Illinois: Illini Union Bookstore, 1962. 3rd ed., 135 pp. Over 1000 entries, including general science, physics, engineering science. 4th ed., 1965, 143 pp.

25 Shipman, Joseph C., "General Science." Library Trends, 15:793-815, April 1967. Discussion of guides, bibliographies, and major reference materials, including foreign titles. 91 refs. See 6.

26 Phelps, Ralph H., "Engineering." Library Trends, 15:868-879, April 1967. Discussion of the bibliographical services available to engineers. General and specialized engineering services, library services, and guides to the literature. 21 refs. See 6.

27 Grogan, Denis J., Science and Technology. An Introduction to the Literature. London: Clive Bingley; Hamden, Conn.: Archon Books, 1970. 231 pp. A textual survey and guide to the literature by type, not subject. General reference materials, indexes and abstracts, services, conference proceedings, and translations are samples of the 20 chapters. Patents, standards, nontextual materials and trade literature are included. The index does not include titles mentioned in the text.

D. Bibliographies and Guides: Histories of Science and Technology

[Isis] (29) is the major source for reviews on historical scientific literature from 1913 (more on science than technology). The Technical Book Review Index (32) serves for general reviews from 1917, but material is difficult to find because subjects are not classified. For material of recent date, Technology and Culture (37) is an excellent source for reviews and book titles related to the history of technology.

Reviews, notices and lists of current books are featured in most periodicals and usually entered collectively in the volume indexes. Unfortunately, these potentially useful items are scattered throughout the literature and are difficult to recover. Bulletins, pamphlets, publishers announcements, manufacturers catalogs and similar ephemeral publications (along with news items) are even more elusive and seldom recorded in bulk references.

28 [John Crerar Library]. A List of Books on the History of Science. Chicago: The Library, 1911. 297 pp. Prepared by Aksel G. S. Josephson. 1500 classified entries, some with annotations. Supplement, 1917, 139 pp., 800 items. 2nd supplement, 1942-1946, prepared by R. B. Gordon. Includes physics and general science, but not engineering.

29 [Isis]. "Critical Bibliography of the History of Science and its Cultural Influences." From 1913, currently (the 94th) in a 5th annual issue; Vol. 60, Pt. 5, 1969, 219 pp. 3510 books and articles arranged alphabetically by author in four main groups and 112 subdivisions. General refs., pp. 13-18. Technology, pp. 58-65 (822-958). Technology for 19th and 20th centuries, pp. 174-178 (3029-3113); from 1914, pp. 193-196 (3423-3467). Name index, pp. 199-219. Vol. 21 (1934), pp. 502-698, contains an index and table of contents for the first 20 volumes.

30 [John Crerar Library]. A List of Books on the History of Industry and Industrial Arts. Chicago: The Library, 1915. Reprinted, Detroit: Gale, 1966. 486 pp. Prepared by Aksel G. S. Josephson. About 3300 titles classified by subject. History of electrical engineering, pp. 60-68, 57 entries, of which 34 are in English. Although limited to library holdings, this is a useful list of early works, with numerous pertinent items in other categories.

31 [John Crerar Library]. Subject Bibliography to Histories of Scientific and Technical Subjects. Chicago: The Library, 1933-1944, 3 vols.

32 Technical Book Review Index. New York: Special Libraries Association. First published 1917. Current

series, Vol. 1, 1935-1936. Compiled by the Science and Technology Department, Carnegie Library of Pittsburgh. Author entry, book title, publisher and address, date, number of pages, price. Each entry consists of one or more reviews, or extracts, with source and other details. Annual author index, but no guide to subjects.

33 [Cooper Union Library]. A Guide to the Literature on the History of Engineering, Available in The Cooper Union Library. New York: Cooper Union, 1946. 46 pp. Engineering and Science Series, Bulletin 28. Reprinted with additions, 1947. 665 classified items.

34 Higgins, Thomas J., "A Classified Bibliography of Publications on the History and Development of Electrical Engineering and Electrophysics." Bulletin of Bibliography, 20:67-71, 85-92, 115-122, 136-142, 160-162, Sept.-Dec. 1950, Jan.-April 1952. Also: University of Wisconsin, Reprint No. 198. About 1200 books and articles classified under 23 headings. This comprehensive survey includes all aspects of power engineering, the telecommunications fields, corporations and societies.

35 Sarton, George, Horus: A Guide to the History of Science. A first guide for the study of the history of science with introductory essays on science and tradition. Waltham, Mass.: Chronica Botanica Co., 1952. 316 pp. Pt. 1, pp. 3-66, Introductory essays. Pt. 2, pp. 69-303, is a select bibliography in four sections; history, science, history of science, and organization of the study and teaching of the history of science, with specific topics grouped in 26 chapters. Covers world-wide literature; books, journals, reference sources, also societies, institutions, museums, libraries and international congresses. Some chapters are arranged by countries, cultural groups, or periods. Chap. 19, pp. 149-193, has 41 grouped entries on the history of special sciences. These include physics, electricity and magnetism, and technology, "inventions." There are 39 items in English in these 3 groups. A basic book for the serious student.

36 Poole, Mary E. (comp.), "History" References from the Industrial Arts Index, 1913-1957. Raleigh, North Carolina: D. H. Hill Library, N. C. State College, 1958. 119 pp. Over 3000 entries from periodicals, classified by subject. About 300 items refer to electricity, electrical engineering, telegraph, telephone, radio, television. Many

articles of historical interest are not listed. See 88.

37 [Technology and Culture]. "Current Bibliography in the History of Technology." First annual list (of works published in 1962), Vol. 5, pp. 138-148, 1964. Alphabetically arranged by author in six chronological divisions and 15 categories. These include general and collected works, documentation, biography, mechanical and electromechanical technology, and communications and records (telegraphy, telephony, radio, the phonograph and photography). Brief descriptions, mention of a review source, and excerpts from reviews are given with some entries. Author index. Cumulative subject index for the first four lists (1962-1965), Vol. 8, pp. 291-309, Spring, 1968. Compiled by Jack Goodwin.

38 Rider, K. J., The History of Science and Technology. A Select Bibliography for Students. London: Library Association, 1967. 60 pp. A brief, annotated list for students and the general reader.

39 Neu, John, "The History of Science." Library Trends, 15:776-792, April 1967. A survey of bibliographies (books and articles) on the history of science and technology. 49 refs. See 6.

40 Ferguson, Eugene S., Bibliography of the History of Technology. Cambridge, Mass.: The M.I.T. Press and The Society for the History of Technology, 1968. 347 pp. There are 5 chapters on general reference works. Four separate chapters deal with biography, government publications, periodicals, and technical societies. One long chapter (pp. 213-306), covers 12 subject fields. The section on electrical and electronic arts has 35 entries. Other chapters list early source books, manuscripts, illustrations, travel and description, technology and culture. Most entries are annotated, frequently with critical as well as helpful comments. Numerous foreign-language works, but English titles not given. Introductory comments with supplemental information are given with most chapters and some sections. There are many cross references and a full index is provided. A basic guide for the student and technical historian.

E. Dictionaries, Encyclopedias and Handbooks

The various forms of general condensations, or compendia, grouped in this section cover a broad range of topics. Only a few representative titles of historical interest are given. Comparison of successive editions--particularly in regard to the expansion of contents, appearance of new topics and the deletion of obsolete material--is a good way to assess the growth of knowledge and changes in contemporary practices.

Earlier works are usually more general and lack references. Later editions, especially those published after World War I, reveal the trend toward specialization and more thorough documentation. Similarly, earlier works are largely one-man compilations, whereas later editions are group contributions by specialists. Other subject handbooks, annuals, and yearbooks restricted to specific fields are listed elsewhere in the respective sections.

41 Rodwell, George F. (ed.), A Dictionary of Science.... London: Ward, Lock and Tyler, 1871. 580 pp. Philadelphia: Henry C. Lea, 1873. 694 pp. "Preceded by an essay on the history of the physical sciences." The wide-ranging contents include electricity and magnetism.

42 Knight, Edward H., Knight's American Mechanical Dictionary. New York: J. B. Ford, 1874-1876, 3 vols. 2831 pp., about 5000 illus. Also, New Mechanical Dictionary (Boston: 1884), 968 pp., 2605 illus. This is a supplementary volume in 4 parts. Various other editions and publishers. A panoramic view of contemporary practice; tools, instruments, machines, processes, engineering, appliances, vocabulary, inventions, history, and technology in the sciences and arts.

43 Houston, Edwin J., A Dictionary of Electrical Words, Terms and Phrases. New York: W. J. Johnston, 1889. 655 pp., 396 illus.

44 The Cyclopedia of Electrical Engineering. Philadelphia: Gebbie, 1891. Described as a compilation from the most recent and best works. Contains "a history of the discovery and application of electricity, with its practice and achievement from the earliest period to the present time." Illus.

45 Sloane, T. O'Conor, *The Standard Electrical Dictionary*. New York: Munn, 1892, 624 pp. 2nd ed., 1897, 682 pp., 393 illus.

46 Fowler, William H. (comp.), *Fowler's Electrical Engineer's Pocket Book*. Manchester (London): Scientific Publishing Co. Annual from 1900. Diags., tables. 1964, 590 pp.

47 *Whittaker's Electrical Engineer's Pocket-Book*. London, New York: Whittaker, 1903. Kenelm Edgcumbe (ed.), 456 pp. 3rd ed., 1911, 596 pp. London: Pitman, 1920. Reginald E. Neale (ed.), 4th ed., 682 pp.

48 Knowlton, Archer E. (ed.), *Standard Handbook for Electrical Engineers*. New York: McGraw-Hill, 1907. Other eds: 5th, 1922; 6th, 1933; 7th, 1941; 8th, 1948. 9th ed., 1957, 2230 pp., 1725 illus. 10th ed., 1968, Donald G. Fink and John M. Carroll (eds.).

49 Pender, Harold (ed.), *Handbook for Electrical Engineers*. New York: John Wiley, 1914. H. Pender and William A. Del Mar (eds.), 2nd ed., 1922. 3rd ed., 1936, 2 vols., *American Handbook for Electrical Engineers*. 4th ed., 1949-1950, 2 vols., *Electrical Engineers' Handbook*. Vol. 1, Electric power, 1716 pp. Vol. 2, Electrical communications and electronics, H. Pender and Knox McIlwain (eds.), 1618 pp. This vol. has 23 sections by 78 contributors. There are extensive bibliographies.

50 Molesworth, Walter H. (ed.), *Spon's Electrical Pocket-Book*. London: E. and F.N. Spon; New York: Spon and Chamberlain, 1916. 488 pp., 325 illus., diags. "A reference book of general electrical information, formulae and tables for practical engineers." 4th ed., 1932. 8th ed., 1941, and later annual eds.

51 Glazebrook, Richard T. (ed.), *A Dictionary of Applied Physics*. London: Macmillan, 1922. 5 vols., reprinted, New York: Peter Smith, 1950. Vol. 2, Electricity, 1104 pp., 56 articles by 41 contributors in addition to the usual dictionary entries. Topics of particular interest; Thermionic valves, Thermionics, Electron theory, Electrons and the discharge tube, Photoelectricity, X rays, Telegraph, Telephony, Wireless telegraphy, Radio frequency measurements. Numerous illustrations and cross references. Some articles have footnotes, others contain a list of references.

A standard work with much of historical value.

52 Collins, A. Frederick, The Radio Amateur's Handbook. New York: Thomas Y. Crowell, 1922. 8th ed., E. L. Bragdon (rev.), 1941. 341 pp., 118 photos, diags.

53 Radio Handbook. Philadelphia: J. C. Winston, 1923. Dart, Harry F. (comp.), and Francis H. Doane (ed.), International Correspondence Schools. 514 pp., diags., charts, tables.

54 The Radio Amateur's Handbook. West Hartford, Conn.: American Radio Relay League. Annually from 1926. Covers all aspects of radio of special interest to amateurs and constructors, with numerous charts, tables, diagrams and photos. This standard manual includes instructional text and detailed descriptions for building and operating equipment as well as a wide range of reference material. Each issue is revised to include current practices.

55 Gernsback, Sidney, S. Gernsback's Radio Encyclopedia. New York: Experimenter Publishing Co., 1927. 168 pp., illus. 2nd ed., 1931. 352 pp., illus.

56 Manly, Harold P. (comp.), Drake's Cyclopedia of Radio. Chicago: F. J. Drake, 1927, 869 pp. 5th ed., 1932, 1045 pp. 10th ed., 1942, Drake's Cyclopedia of Radio and Electronics, H. P. Manly and L. O. Gorder. A practical reference book; illus., charts, diags., tables. Covers radio, audio systems, photocells, television, etc. Later eds.

57 The B. B. C. Handbook. London: British Broadcasting Corporation. Annually from 1928. 1929 ed., 480 pp. The 4 sections include one on technical matters, pp. 281-403; and one on references, mainly technical, pp. 406-458. Numerous photos, diagrams, formulas, and a glossary. Recent edition (1970), 303 pp., television, radio, programme services, external services, engineering, reference. The last section contains a list of BBC dates, pp. 246-256, 255 entries with day and month from Nov. 1, 1922 to Dec. 1, 1969. The engineering section, pp. 119-162, contains 21 maps, with lists, showing detailed regional television and radio services.

58 Henney, Keith (ed.), The Radio Engineering Handbook. New York: McGraw-Hill, 1933. 583 pp., copiously

illus., well documented with footnotes and bibliographies. 2nd ed., 1935, 850 pp. 3rd ed., 1941, 945 pp. 4th ed., 1950, 1197 pp., nearly 1050 illus., over 100 tables and numerous charts. The 22 chapters cover components, circuits, measurements, power supplies, amplifiers, oscillators, modulation and detection, propagation, antennas, waveguides, electron tubes, loud-speakers and room acoustics, receiving systems, broadcasting, television, facsimile, navigational aids, code transmission and reception. 5th ed., enlarged and revised with additional chapters, 1959. 1800 pp., 1636 illus.

59 Langford-Smith, Fritz (ed.), Radiotron Designer's Handbook. Sydney, Australia: Wireless Press; Harrison, New Jersey: RCA; also London: Iliffe (Radio Designer's Handbook), 1934, 1935, 1940 (352 pp.). 4th ed., 1952, 1498 pp., over 950 illus., more than 3300 entries in chapter bibliographies and refs., over 100 tables. This comprehensive handbook comprises 7 parts and 38 chapters that thoroughly treat electron tubes (radio valves), general theory, components, audio and radio frequencies, power supplies, receivers. Wide variety of useful data; standards, codes, units, symbols, etc.

60 The Radio Handbook. San Francisco, Santa Barbara: Radio, Ltd., 1935. By editors of Radio (F. C. Jones), (W. W. Smith), (R. L. Dawley), and others. 3rd ed., 1937; 4th, 5th ed., 1938; 6th ed., 1939, 640 pp. 13th ed., 1951 (Editors and Engineers, Ltd.). Photos, diags., tables, charts.

61 The Amateur Radio Handbook. London: Radio Society of Great Britain, 1938. 3rd ed., 1961. 552 pp., illus., diags., tables.

62 Kohlhaas, H. T. (ed.), Reference Data for Radio Engineers. New York: Federal Telephone and Radio Corporation, 1943. 200 pp. Tables, charts, diags. Graphs for general engineering, materials, audio and radio design, propagation and antennas, mathematics. 2nd ed., 1946, 3rd ed., 1949. 4th ed. (H. P. Westman, ed.), New York: IT&T, 1956. 1121 pp., plus 29 p. index. Footnote refs. 5th ed., 1970, 1196 pp., plus 41 p. index, 1350 illus.

63 Cooke, Nelson M., and John Markus, Electronics Dictionary. New York: McGraw-Hill, 1945. 433 pp.,

well illus. A glossary of over 6000 terms used in radio, television, industrial electronics, communications, facsimile, sound recording. New ed., Electronics and Nucleonics Dictionary, 1960. 543 pp., well illus. Over 13,000 terms in all fields of electronics and nucleonics, with abbreviations, definitions and synonyms according to technical standards. Fully cross-referenced. 3rd ed., 1966, John Markus (ed.), 650 pp., about 1000 illus.

64 Kempner, Stanley, Television Encyclopedia. New York: Fairchild, 1948. 415 pp., 82 illus. Pt. 1, Milestones to present-day television, pp. 3-42. Pt. 2, Pioneers and contemporaries in television, pp. 43-135. Pt. 3, Television's technical vocabulary, pp. 137-392. Each part arranged alphabetically. App., The urban market for television, pp. 393-399. Bibliography, pp. 401-415, books, periodicals, pamphlets and papers, about 250 items. Generally useful for secondary sources, light technical treatment, little documentation. No index.

65 Reyner, J. H., The Encyclopedia of Radio and Television. London: Odhams, 1950. 2nd ed., 1957. 736 pp., 800 illus., diags. About 5000 entries arranged alphabetically.

66 Ballantyne, Denis W. G., and Louis E. Q. Walker. A Dictionary of Named Effects and Laws in Chemistry, Physics and Mathematics. London: Chapman and Hall, 1958; New York: Macmillan, 1959. 205 pp., some diagrams. Brief alphabetical entries on principles, laws, formulas, theories, rules, effects, texts, processes, etc. Over 1000 items. 2nd ed., 1961, 234 pp.

67 Sarbacher, Robert I., Encyclopedic Dictionary of Electronics and Nuclear Engineering. Englewood Cliffs, New Jersey: Prentice-Hall, 1959. 1417 pp., about 1400 line illus., tables. Comprehensive reference work with over 14,000 entries and extensive cross references. Standard definitions are used and their sources are given.

68 Süsskind, Charles (ed.), The Encyclopedia of Electronics. New York: Reinhold, 1962. 974 pp., illus. Over 500 articles, primarily of contemporary interest. Short biographies of notable men are included. Little historical matter.

69 Encyclopedia of Science and Technology. New York:

McGraw-Hill, 1966. 15 vols. 9900 pp., 9500 illus., 7400 articles. Supplemented by the Year Book of Science and Technology.

2. SERIAL PUBLICATIONS

A. Guides to Periodicals

70 Bolton, Henry C., A Catalogue of Scientific and Technical Periodicals 1665-1895. Together with Chronological Tables and a Library Check-List. Washington: Smithsonian Institution, 1885. 2nd ed., 1897, reprinted, Johnson, 1965. 1247 pp. 8603 entries of serials published in several languages, foreign titles in the original. Chronological tables, pp. 1017-1123; index, pp. 1125-1140. The analytical subject index, pp. 1141-1154, has the following pertinent entries: electricity, 95; inventions, 85; physics, 105; telegraph and telephone, 29. Also helpful as a guide to related fields.

71 Ulrich's International Periodicals Directory. A Classified Guide to Current Periodicals, Foreign and Domestic. New York: R. R. Bowker. First published 1932. Current ed. (13th), 1969. Two vols., A-L, M-Z, 1659 pp. (double column). Alphabetical by subject and title. Title and subject index, pp. 1414-1659. Pertinent sections include: Communications (general, radio and television, telephone and telegraph), pp. 364-375; Electricity and electrical engineering, pp. 474-489; Science, pp. 1214-1234. Entries give various details; first year of publication, publisher and address, frequency of issue, price, circulation, index sources, and brief contents, such as abstracts, bibliographies, book reviews, illustrations, charts, index.

72 Smith, William A., Francis L. Kent (eds.), assisted by George B. Stratton, World List of Scientific Periodicals Published in the Years 1900-1950. London: Butterworth; New York: Academic Press, 1952, 3rd ed., 1058 pp. About 50,000 titles in 24,029 numbered entries. 4th ed. (Peter Brown and George B. Stratton, eds.), 3 vols., 1963-1965, extends the list to 1960. There is a list of periodic international congresses in vol. 3, pp. 1789-1824. Gives British locations.

73 England, Rosemary, Union List of Periodicals on Electronics and Related Subjects. London: Aslib, 1961. 62 pp. About 1250 entries.

74 Fowler, Maureen J., Guides to Scientific Periodicals. Annotated Bibliography. London: Library Association, 1966. 318 pp., 1048 entries. Universal guides, pp. 1-78. National and regional guides, pp. 83-269.

B. Guides to General Periodical Literature

75 Poole, William F., Poole's Index to Periodical Literature. Boston: Houghton, Mifflin. Vol. 1, 1802-1881, in 2 parts, A-J, K-Z, 1440 pp., revised ed., 1893. The seven vols. have about 50 columns related to electricity, magnetism, telegraphy and telephony, including wireless. Reprinted, New York: Peter Smith, 1938. Five supplementary vols. span the years 1882 to 1906.

76 Nineteenth Century Readers' Guide to Periodical Literature. New York: H. W. Wilson, 1944. In 2 vols., A-K, L-Z, 3074 pp. Covers 1890-1899, with supplementary indexing 1900-1922. Subject classification with cross references.

77 Readers' Guide to Periodical Literature. A consolidation of the cumulative index to a select list of periodicals and the readers' guide to periodical literature. New York: H. W. Wilson. Author and subject index. Vol. 1, 1900-1904, 1640 pp., multiple annual cumulations thereafter. Subject classification with topical breakdown. In Vol. 1, electricity, etc., 20 columns; telegraphy and the telephone, 4 columns; wireless telegraphy, 3 columns, with a few items on wireless telephony. Includes early material from the following publications: Cassier's, Century, Engineering Magazine, Living Age, Popular Science Monthly, World's Work, Science, Smithsonian Institution Annual Report; and later, Scientific Monthly, Current Biography, Fortune, Radio Broadcast, Radio and Television News.

78 International Index to Periodicals. New York: H. W. Wilson. Author and subject index to a selected list of periodicals of the world, devoted chiefly to the humanities and science. Formerly Readers' Guide Supplement. Vol. 1, 1907-1915, multiple annual cumulations thereafter. Title changed to Social Sciences and Humanities Index with Vol.

19 (1965-66). Includes such publications as Discovery, Nature, Proceedings of the National Academy of Sciences, and the Philosophical Transactions of the Royal Society. Very light on technology, but does list entries under electricity, radio, telegraph and telephone.

C. Guides to Scientific Periodical Literature

79 [Royal Society]. Catalogue of Scientific Papers. London: From 1867. Vols. 1-6, 1800-1863; 7, 8, 1864-1873; 9-11, 1874-1883; 12, 1800-1883 supplement; 13-19, 1884-1900. Reprinted, Metuchen, N. J.: Scarecrow, 1967. 23 vols. in 22. Includes all papers published in transactions, journals and related periodicals. Entries, by author's name, give article title, journal, date, volume and pagination.

80 [Royal Society]. Subject Index 1800-1900. Cambridge: University Press, 1908-1914, 3 vols. in 4. 1, Pure Mathematics; 2, Mechanics; 3, Pt. 1, Physics; Pt. 2, Electricity and Magnetism. Vol. 3 contains a list of serial publications, pp. xi-lxxxix. Subject index, pp. 1-550, 551-927. Both parts have a schedule of classification, with page numbers, and an index.

81 [Royal Society]. Index to the Proceedings, 1800-1905. London: 1913. Continued with various titles. Index to Proceedings 1905-1930 and Philosophical Transactions 1901-1930, 1932. Index to Proceedings, Philosophical Transactions and Obituary Notices 1931-1940, also 1941-1950. Index to Proceedings, Philosophical Transactions and Biographical Memoirs 1951-1960. See also 346, 349.

82 [Royal Society]. International Catalogue of Scientific Literature. London: 1902-1919. Originally 254 vols. (17 sections). Reprinted, New York: Johnson, 1968, 32 vols. with 4 pp. on one page. Reprinted, Metuchen, N. J.: Mini-Print Corp., 1969-70, 33 vols. with 4 pp. on one page. Section C, Physics, 2 vols., 1901-1907, 1908-1914. Comprises schedules and indexes in four languages, author's catalogue, subject catalogue, and list of indexed journals with abbreviations.

83 Science Abstracts. Physics and Electrical Engineering. London: Institution of Electrical Engineers and The Physical Society of London. Vol. 1, Jan.-Dec. 1898,

834 pp., 1423 items. Numbered entries classified by subjects in 10 sections per month. There are entries under discharge tubes, radiation, waves, oscillators, telegraph, telephone, in addition to general physics (heat, light, sound, electricity) and general electrical engineering. Each entry summarizes the article, gives particulars of the source, and is signed by the contributor. All titles given in English. Annual name index (including non-author references), subject index, list of journals and list of contributors. Divided in 1903 into two parts: A, Physics, B, Electrical Engineering. Currently in three sections: A, Physics Abstracts; B, Electrical and Electronics Abstracts; C, Control Abstracts. Section A, Vol. 73 (1970), contains 79,830 items; with indexes for patents, reports, conferences, books and bibliographies. *See* 85.

D. Guides to Engineering Periodical Literature

84 Engineering Index. New York: Various publishers from 1884, currently Engineering Index, Inc. Annual issues from 1928 (reprinted by Johnson). 1928 in 2 vols., A-I, J-Z, 2091 pp. About 50,000 items from about 1700 publications in 18 languages. Entries arranged alphabetically by specific subject with subtopics and cross references. Most entries are briefly described by contents. Single annual vol. from 1931 to 1962, 2 vols. per year to date. Recent issue (1969), 3435 pp., with entries from 2205 periodicals. Some items were separately classified under "History" until 1958. Author index.

85 Electrical Engineering Abstracts. London: Institution of Electrical Engineers. Section B of Science Abstracts from 1903. Monthly, with annual cumulations. Annual author and subject index. Numbered entries, contributed items signed, classified by U.D.C. The growth of the literature is indicated by the following listings: Vol. 49, 1946, 2977 items; Vol. 66, 1963, 16,300 items; Vol. 71, 1968, 30,438 items. Over one-third of the entries for recent years are concerned with electronics, telecommunications, and related topics. *See* 83.

86 [Institute of Radio Engineers]. Index to Proceedings 1909-1926. An inclusive index: titles, pp. 5-23; authors, pp. 23-26; cross index, pp. 27-32. Includes index for Wireless Institute. Also for 1909-1936.

87 [Institute of Radio Engineers]. Index to Proceedings 1913-1942. Cumulative index, Vols. 1-30: titles, pp. 3-40; authors, pp. 41-46; subjects, pp. 47-65, nontechnical, pp. 66-69. Also Proc. IRE., Vol. 31, Pt. 2, June 1943.

88 Applied Science and Technology Index. New York: H. W. Wilson. Was Industrial Arts Index until 1958. Subject index to a selected list of engineering and trade periodicals. First annual cumulation for 1913, published 1914, 326 pp. This issue indexed 48 periodicals, including AIEE Transactions, Franklin Institute Journal, Electrical World, Scientific American and the Supplement, and the Bulletin of the U. S. Bureau of Standards. There are entries for electricity, etc., the telegraph, telephone, and wireless. The 1969 vol., 1456 pp., has entries from 230 periodicals published in the United States, Great Britain, and Canada. About 35 periodicals listed are directly or closely concerned with electricity and electronics, and another dozen are of related scientific interest. The subject fields are highly divided, there are full cross references to related topics with most groups, and numerous cross entries. "History" subheadings are provided for specific items, but these do not display all entries that are of historical interest. Entries are by titles; some indicate illus., diags., maps and occasionally a bibliography. Some books are included. See 36.

89 Maynard, Katherine, (ed.), A Bibliography of Bibliographies in Electrical Engineering, 1918-1929. Providence, R.I.: Special Libraries Association, 1931. 156 pp., 2250 references to articles containing bibliographic entries. Useful guide to papers, including French and German, for the period indicated.

90 RCA Technical Papers, Index. Princeton, N. J.: RCA Review. Vol. 1, 1919-1945, 1-1778. Vol. 2, 1946-1950, 1779-2801. Vol. 3, 1951-1955, 2801a-3864. Vol. 4, 1956-1960, 3865-5227. Each volume is divided into chronological, alphabetical, author and subject indexes. Books and articles in the English language by Radio Corporation of America personnel.

91 [American Institute of Electrical Engineers]. Index to Transactions. Vol. 41 to 57 (1922-1938), 1939. Subjects, pp. 5-105; authors, pp. 107-156. Alphabetical with multiple cross references. Technical papers and discussions. Vol. 58 to 68 (1939-1949), 1950. Subjects, pp. 3-

147; authors, pp. 148-200.

92 Petraglia, Frank A., (ed.), Electronic Engineering Master Index. A subject index to electronic engineering periodicals, January 1925 to June 1945. New York: Electronics Research Publishing Co., 1945. 318 pp. About 15,000 entries from 65 publications, arranged alphabetically. Pt. 1, Jan. 1925 through Dec. 1934, pp. 1-108; Pt. 2, Jan. 1935 to June 1945, pp. 109-310. Subject index, pp. 311-318. No author index. Full page numbers not given. History references, pp. 42, 43, 36 items; pp. 184, 185, 21 items. Later supplements also published.

93 Rettenmeyer, Francis X., "Radio-Electronic Bibliography." Radio, May 1942 to Aug. 1945. (Also titled "Radio Bibliography.") A comprehensive listing of world periodicals, with some textbooks. All titles given in English. The series contains about 8000 items arranged in the following 19 categories. 1, Aviation radio. 2, Frequency modulation. 3, Crystallography. 4, Tubes. 5, Amplification, detection, oscillators, coils. 6, Filters, sound, loudspeakers. 7, Remote control. 8, Antennas and radiation. 9, Propagation. 10, Measurements. 11, Direction finding. 12, Television. 13, Cathode-ray oscillographs. 14, Interference and Static. 15, Household receivers. 16, Transmitters. 17, Aircraft radio. 18, Velocity modulation. 19, Frequency standards (by National Bureau of Standards). These are listed elsewhere by subject.

94 [Institute of Radio Engineers]. Index to Proceedings 1943-1947. Proc. IRE., Vol. 36, Pt. 2, June 1948.

95 [Institute of Radio Engineers]. Index to Abstracts and References 1946-1953. Proc. IRE., Vol. 43, Pt. 2, Nov. 1955. 189 pp. 8 yearly sections, author and subject index and list of journals.

96 Dorf, Richard H., Radiofile. New York: Radio Magazines, 1946-1955. Bimonthly and annual issues. An index of articles in leading radio magazines. Listings cross-indexed and identified by title, magazine, month and page number. Became Lectrodex in 1956.

97 [Institute of Radio Engineers]. Cumulative Index of IRE Publications 1948-1953. Also, 1954-1958.

E. Magazines and Journals

The following list contains almost all of the serial publications mentioned in this bibliography and a few additional titles likely to be useful to the reader. They are arranged alphabetically with the abbreviated form or title as used in the subject entries. Although no systematic effort has been made to record every title change or cessation, some specific titles and the periods covered by them have been given for certain publications most frequently cited. With a few exceptions, new publications of recent origin (1950 or later) are not included.

98 Am. J. Phys. American Journal of Physics. New York: American Institute of Physics, 1933.

99 Am. J. Sci. American Journal of Science. New Haven, Conn., 1818. From 1820 to 1879, American Journal of Science and Arts (Silliman's Journal...).

100 Ann. Am. Acad. Annals of the American Academy of Political and Social Science. Philadelphia, 1890.

101 Ann. of Sci. Annals of Science. A quarterly review of the history of science and technology since the Renaissance. London, 1936.

102 Audio. San Francisco, Santa Barbara, Philadelphia, 1917. Various titles: Pacific Radio News, 1917-1921; Radio, 1921-1947; Audio Engineering, 1947-1954.

Audio Engineering. See Audio.

103 Bell Lab. Rec. Bell Laboratories Record. New York, 1925.

104 B. S. T. J. Bell System Technical Journal. New York, 1922.

105 Bell Tel. Mag. Bell Telephone Magazine. New York, 1922. Was Bell Telephone Quarterly, 1922-1941.

Bell Tel. Qtly. See Bell Tel. Mag.

106 Brit. J. App. Phys. British Journal of Applied

Physics. London: Institute of Physics, 1950. Superseded by Journal of Physics, Section D, 1968.

107 Brit. J. Hist. Sci. The British Journal for the History of Science. Keston, Kent: The British Society for the History of Science, 1962.

108 Broadcasting. Washington, 1931. Variations in title.

109 Cassier's Mag. Cassier's Magazine. New York, 1891-1913.

Communication and Broadcast Engineering. See Communications.

110 Communication and Electronics. New York: American Institute of Electrical Engineers, July 1952 to Nov. 1963.

111 Communication Engineering. Great Barrington, Mass., 1940-1954. 1944-1948, FM and Television.

112 Communications. New York, 1937. Incorporating Radio Engineering, 1920, and Communication and Broadcast Engineering, 1934. Superseded by Television Engineering, then F.M. Magazine.

113 Discovery. London, 1920-1966. Merged with Science Journal.

114 Electric J. Electric Journal. Pittsburgh: Electric Club, 1904.

115 Electrical J. Electrical Journal. London, 1878.

116 Elect. Comm. Electrical Communication. New York: International Telephone and Telegraph Corporation, 1922.

117 Elect. Eng. Electrical Engineering. New York: American Institute of Electrical Engineers, 1884. Transactions from Vol. 1. Proceedings, 1905-1919. Journal, 1920-1930. Transactions also 1905-1933. Separate section issued as Transactions, 1934. Last issue Dec. 1963. See also IEE Spectrum, Proc. IEEE.

118 Elect. Rev. Electrical Review. London, 1872. Was

Telegraphic Journal and Electrical Review, 1873-1891.

119 Electrical Times. London, 1891.

120 Electrical World. New York, 1874.

121 Electrician. London, 1861. Incorporated in Electrical Review. Name changed to The Electrical Journal (Vol. 149) Oct. 17, 1952.

122 Electronic Age. New York: Radio Corporation of America, 1941. Vol. 1 to 16, Radio Age.

123 Electronic Engineering. London, 1928. Various titles: Television (1-8), 1928-1935; Television and Short-Wave World (8-12), 1935-1939; Electronics and Television and Short-Wave World (12-14), 1939-1941.

124 Electronic Industries. Philadelphia: Caldwell-Clements, 1942. Various titles; Tele-Tech, Tele-Tech and Electronic Industries, Electronic Industries and Tele-Tech. Currently Electronic Engineer.

Electronic and Radio Engineer; see Electronic Technology.

125 Electronic Technology. London, 1923. Various titles: Experimental Wireless, 1923-1931; Wireless Engineer, 1931-1935; also Electronic and Radio Engineer. Superseded by Industrial Electronics, 1962.

126 Electronics. New York: McGraw-Hill, 1930.

Electronics and Television and Short-Wave World; see Electronic Engineering.

127 Electronics World. New York, 1919. Various titles: 1919-1920, Radio Amateur News; 1920-1948, Radio News.

128 Endeavour. London: Imperial Chemical Industries, 1942.

129 Engineer. London, 1856.

130 Engineering. London, 1866.

131 English Mechanic. London, 1865-1926.

Experimental Wireless; see Electronic Technology.

FM and Television; see Communication Engineering.

132 Fortune. New York, 1930.

133 G. E. Rev. General Electric Review. Schenectady, N. Y.: General Electric Company, 1903.

134 IEEE Spectrum. New York: Institute of Electrical and Electronics Engineers, 1964. See also Elect. Eng.

135 Int. TV Tech. Rev. International TV Technical Review. London, 1959. Incorporated in International Broadcast Engineer, 1964.

136 Isis. An International Review devoted to the History of Science and Its Cultural Influences. Brussels, 1913. Official journal of the History of Science Society, George Sarton (ed.). Various publishing offices and editors. From June 1964, Robert P. Multhauf (ed.), Smithsonian Institution.

J. AIEE. See Elect. Eng.

137 J. App. Phys. Journal of Applied Physics. Lancaster, Pa.; New York: American Institute of Physics, 1931. Formerly Physics.

138 J. Brit. IRE. Journal of the British Institution of Radio Engineers. London, 1926. See also Proc. IERE., and Radio and Electronic Engineer.

139 J. Broadcasting. Journal of Broadcasting. Philadelphia, 1956.

140 J. Eng. Ed. Journal of Engineering Education. Lancaster, Pa.; Washington: Society for the Promotion of Engineering Education, 1910.

141 J. F. I. Journal of The Franklin Institute. Philadelphia, 1826.

142 J. IEE. Journal of the Institution of Electrical Engi-

neers. London: (Society of Telegraph Engineers) 1872. From 1941: Pt. 1, General; Pt. 2, Power engineering; Pt. 3, Radio and communication engineering. (New series, Jan. 1955.) See also Proc. IEE.

Journal of Physics. See Proc. Phys. Soc.

143 J. POEE. Post Office Electrical Engineers' Journal. London: Institution of Post Office Electrical Engineers, 1908.

144 J. Roy. Soc. Arts. Journal of the Royal Society of Arts, Manufactures, and Commerce. London, 1852.

145 J. Sci. Insts. Journal of Scientific Instruments. London: Institute of Physics, 1923. Currently, Journal of Physics, Section E: Scientific Instruments. Institute of Physics and Physical Society.

J. SMPE. Journal of the Society of Motion Picture Engineers. See J. SMPTE.

146 J. SMPTE. Journal of the Society of Motion Picture and Television Engineers. Washington (New York), 1916. Was Society of Motion Picture Engineers to 1949.

147 J. Television Soc. Journal of the Television Society. London, 1930.

148 Marconigraph. London, 1911-1913. American edition became Wireless Age. English edition became Wireless World.

149 Midwest Engineer. Evanston, Ill.: Western Society of Engineers, Chicago, 1948.

150 Nature. London, 1869.

Pacific Radio News. See Audio.

151 Phil. Mag. Philosophical Magazine. London, 1798. Various titles; London, Edinburgh and Dublin Philosophical Magazine and Journal of Science, 1840-1948. Numerous series.

152 Phil. Trans. Philosophical Transactions of the Royal Society. London, 1665. Series A, Mathematical and

physical papers, 1887.

153 Phys. Rev. The Physical Review. Lancaster, Pa.; Ithaca, N.Y.: The American Physical Society, 1913. Previously Cornell University, 1893.

Physics. See J. App. Phys.

154 Physics Today. New York: American Institute of Physics, 1948.

155 Popular Radio. New York, 1922-1928. Absorbed Wireless Age.

156 Pop. Sci. Mon. Popular Science Monthly. New York, 1872.

Proc. AIEE. Proceedings of the American Institute of Electrical Engineers. See Elect. Eng.

157 Proc. IEE. Proceedings of the Institution of Electrical Engineers. London, 1872. 1955: Pt. A, Power engineering; Pt. B, Electronics and communications; Pt. C, Monographs. (Proc. IEE., Wireless Section, 1926.) See also J. IEE.

158 Proc. IEEE. Proceedings of the Institute of Electrical and Electronics Engineers. New York, 1913 (1964). Was Proceedings of the Institute of Radio Engineers to Dec. 1963. See also Elect. Eng.

159 Proc. IERE. Proceedings of the Institution of Electronic and Radio Engineers. London, 1962. See also J. Brit. IRE.

Proc. IRE. Proceedings of the Institute of Radio Engineers. New York, 1913. See Proc. IEEE.

160 Proc. IRE Aust. Proceedings of The Institution of Radio Engineers Australia. Sydney, N.S.W., 1937. Became Proc. IREE Aust. (The Institution of Radio and Electronics Engineers Australia), 1964.

161 Proc. Phys. Soc. Proceedings, Physical Society. London, 1874. Issued in two parts from 1949; Pt. A, General physics. Currently Journal of Physics.

162 Proc. Radio Club of America. New York, 1920.

163 Proc. Roy. Inst. Proceedings of the Royal Institution of Great Britain. London, 1851.

164 Proc. Roy. Soc. Proceedings of The Royal Society. London, 1832. Series A, Mathematical and physical sciences, 1866.

165 QST. Devoted entirely to amateur radio. Hartford (Newington), Conn.: American Radio Relay League, 1915. Official organ of the A. R. R. L. and of the International Amateur Radio Union.

Radio. See Audio.

166 Radio Age. Chicago, 1922-1928.

Radio Age. See Electronic Age (RCA).

Radio Amateur News. See Electronics World.

167 Radio Broadcast. Garden City, N. Y., 1922-1930. (Merged with Radio Digest.)

Radio-Craft. See Radio-Electronics.

168 Radio and Electronic Engineer. London: Institution of Electronic and Radio Engineers, 1926. See also J. Brit. IRE.

169 Radio-Electronics. New York, 1929. Various titles: 1929-1943, Radio-Craft; 1943-1948, Radio-Craft and Popular Electronics.

Radio Engineering. See Communications.

Radio News. See Electronics World.

170 Radio Review. London, 1919-1922. See Wireless World.

Radio and Television News. See Electronics World.

171 Radio World. New York, 1922-1939.

172 RCA Rev. RCA Review. New York: Radio Corpora-

tion of America, 1936.

173 Rev. Mod. Phys. Reviews of Modern Physics. New York: American Physical Society, American Institute of Physics, 1929.

174 Rev. Sci. Inst. The Review of Scientific Instruments. New York: American Institute of Physics, 1930.

175 RSGB Bull. RSGB Bulletin. London: Radio Society of Great Britain, 1925. Formerly, T and R Bulletin. The Wireless World was the official journal of the Society until 1925.

176 Science. Washington: American Association for the Advancement of Science, 1880.

177 Sci. Am. Scientific American. New York, 1845. Weekly, became monthly, Nov. 1921.

178 Sci. Am. Supp. Scientific American Supplement. New York, 1876-1919. Was Scientific American Monthly, 1920-1921.

179 Sci. Mon. The Scientific Monthly. Washington: American Association for the Advancement of Science, 1915-1957. Merged into Science.

180 Technology and Culture. The International Quarterly of the Society for the History of Technology. The Society, and University of Chicago Press, 1960.

181 Tech. Rev. Technology Review. Boston: Massachusetts Institute of Technology, 1899.

182 Telegraph and Telephone Age. New York, 1883.

Tele-Tech. See Electronic Industries.

183 Television. London, 1928. Various titles. See Electronic Engineering.

184 Television Quarterly. Syracuse, New York: Journal of the National Academy of Television Arts and Sciences, 1962.

Television and Short-Wave World. See Electronic

Engineering.

Trans. AIEE. Transactions of the American Institute of Electrical Engineers. See Elect. Eng.

185 Trans. IEEE. Transactions of the Institute of Electrical and Electronics Engineers. New York. Professional group publications, titles according to field of interest. Various issues, various years from 1951. Were Trans. IRE until 1964.

Trans. IRE. Transactions of the Institute of Radio Engineers. See Trans. IEEE.

186 Western Electric Engineer. New York: Western Electric Company, 1957.

187 Wireless Age. New York: Marconi Wireless Telegraph Company of America, 1913-1919; Radio Corporation of America, 1919-1925. Merged with Popular Radio. See Marconigraph.

Wireless Engineer. See Electronic Technology.

188 Wireless World. London: The Marconi Press Agency, 1913. Was Marconigraph. Absorbed Radio Review.

3. GENERAL HISTORIES

The entries in this chapter range from specialized works on science and histories of physics to popular books that tell a story about scientific or technological progress. They all cover more topics than any of the comparable works listed elsewhere in the respective subject chapters. The list is highly selective; only those works with some content directly related to electrical science, engineering, and electronics over the period of interest have been included.

A. Physics

189 Ganot, Adolphe (E. Atkinson, trans.), Elementary Treatise on Physics: Experimental and Applied. London: Longmans, Green, 1863. 4th ed., 1870, 888 pp., 698 woodcuts. 13th ed., 1890, 1084 pp., 987 woodcuts and 9 colored plates and maps. Magnetism, pp. 649-687; Frictional electricity, pp. 688-769; Dynamical electricity, pp. 770-994. A standard and popular text with many excellent illustrations of contemporary equipment. Slight historical treatment with some names and dates in the text.

190 Deschanel, A. Privat (J. D. Everett, trans. and ed.), Elementary Treatise on Natural Philosophy. London: Blackie, 1872. First English ed., 1050 pp. New York: D. Appleton, 1876. 1068 pp., 760 wood engrs., 3 color plates. About one quarter of the book is devoted to electricity and magnetism. Occasional footnotes and other references in the text. This standard text of French origin is valuable for the historical treatment, contemporary views, and many fine illustrations. Numerous later editions.

191 Guillemin, Amédée (Mrs. Norman Lockyer, trans., J. Norman Lockyer, ed.), The Applications of Physical Forces. London: Macmillan, 1877. 741 pp., 467 wood engrs., 26 plates, some in color. Magnetism and electricity, pp. 519-726, includes over 100 pages on the electric telegraph. There are other chapters on electric horology, mo-

tors, electric light, electroplating and miscellaneous applications. A practical book with some names and dates in the text, otherwise not documented.

192 Magie, William F., A Source Book in Physics. New York: McGraw-Hill, 1935; Harvard University Press, 1965. 620 pp., 111 diags. This standard book consists of extracts from original works with brief biographical notes and source references. Of the 116 samples, 35 concern magnetism, electricity, conduction in gases, and electromagnetic radiation. Other sections are mechanics, properties of matter, sound, heat and light. From Galileo to the end of the 19th century.

193 Dibner, Bern, Heralds of Science as represented by two hundred epochal books and pamphlets selected from the Burndy Library. Norwalk, Conn.: Burndy Library, 1955. 96 pp., 83 ports., views and facsimiles. Contains 21 items on electricity, magnetism, the telegraph and telephone, from Peregrinus (1558) to Marconi (1897), and others on cathode rays, X rays and the electron. There are brief notes with each entry. (Up to 1900.)

194 Harvey, E. Newton, A History of Luminescence From the Earliest Times until 1900. Philadelphia: American Philosophical Society, 1957. 692 pp., 50 plates. Selected bibliography, pp. 601-669, contains about 1750 items, including foreign works (English titles not given), from 1555 to 1900. Parts of present interest include Chap. 7, pp. 251-304, Electroluminescence, 68 footnotes; Chap. 12, pp. 410-422, Radioluminescence, 22 footnotes. A comprehensive work, thoroughly documented.

195 Sharlin, Harold I., The Convergent Century. The Unification of Science in the Nineteenth Century. London: Abelard-Schuman, 1966. 229 pp. Notes, pp. 207-216, 233 entries arranged by chapters. Bibliography, pp. 217-222, 82 entries. A history limited to electricity, heat, chemistry, and light, with stress on controversies. Chap. 10, pp. 161-181, is on cathode rays and electronics. A useful and interesting survey.

196 Sedgwick, William T., and Harry W. Tyler, A Short History of Science. New York: Macmillan, 1917, 1919, 474 pp. Revised by H. W. Tyler and Robert P. Bigelow, 1939. 512 pp., 80 illus., including plates. App. A, pp. 457-471; Some inventions of the eighteenth and nineteenth

centuries, applied science and engineering, 21 groups. App. B, pp. 472-486; Some important names, dates, and events in the history of science and civilization, from the Egyptian era to 1914. App. C, pp. 487-500; lists 371 items: general books, biographies, source books, journals. There is a short list of references for reading with the chapters. Numerous extracts in the text, with sources. Electrical science in the 18th and 19th centuries is mentioned.

197 Cajori, Florian, A History of Physics in its Elementary Branches. New York: Macmillan, 1899. 330 pp., 18 illus. Reprinted 1924. Also, A History of Physics in its Elementary Branches (through 1925): Including the Evolution of Physical Laboratories. 1929, Dover, 1962. 424 pp., 16 illus. 870 detailed footnote refs. This standard work deals nonmathematically with theories and basic topics from the Greek era up to the 1920s. An essential book for the student and researcher. The last section on the evolution of physical laboratories, pp. 387-406, gives an interesting and useful survey of a few educational and research institutions.

198 Chase, Carl T., A History of Experimental Physics. New York: D. Van Nostrand, 1932. 196 pp., 15 illus. Chronological table of eminent physicists, pp. 185-189, 69 entries, with dates. Much of electronic interest with historical perspective. Nonmathematical, without refs.

199 Pledge, Humphry T., Science Since 1500. A Short History of Mathematics, Physics, Chemistry, Biology. London: H. M. S. O., 1939. 357 pp., 15 plates, 13 graphs, charts and maps. Bibliographical note, pp. 326-329, is a list of books on the history of science grouped by main subjects. From the earliest times up to the modern period. The line diagrams show accuracy of measurements, the connections between master and pupil (scientists), the tracks of science to 1500, and the birthplaces of scientists (with lists).

200 Taylor, Lloyd W., Physics. The Pioneer Science. New York: Houghton, Mifflin, 1941; Dover, 1959, 2 vols. Vol. 1 (394 pp., 211 illus.), mechanics, heat, sound. Vol. 2 (pp. 397-847, illus. 212-551), light, electricity. App. contains a list of 291 book refs. Many footnote refs. to periodicals. A standard work with the subject matter presented in a historical context. Pages 730 to 832 on the telegraph, telephone, radio, decline of the mechanical view,

cathode rays, the electron, X rays, quantum theory and atomic structure are especially useful.

201 Shapley, Harlow, Samuel Rapport, and Helen Wright, A Treasury of Science. New York: Harper and Brothers, 1943. 4th ed., 1958. 776 pp. A well-known compilation of original writings from books and articles. In four groups: science and the scientist, the physical world, the world of life, the world of man. A few pieces of present interest include the atom, electronics, computers.

202 Jaffe, Bernard, Men of Science in America. New York: Simon and Schuster, 1944. 600 pp., 25 illus., 16 plates. A collection of 19 biographical studies subtitled "The role of science in the growth of our country." Three chapters of electrical interest concern Benjamin Franklin, Joseph Henry, and Ernest O. Lawrence. Sources and reference material, pp. 555-571, contain 272 entries listed by chapters.

203 Taylor, F. Sherwood, Science Past and Present. London: William Heinemann, 1945. 275 pp., 42 illus., 22 plates. A broad survey from earliest times with numerous extracts from original works. Reading list, pp. 268-270, 61 items. Chap. 13, pp. 199-208, The age of electricity. Many refs. are given in the text and in frequent footnotes. See 206.

204 Moulton, Forest R., and Justus J. Schifferes (eds.), The Autobiography of Science. Garden City, N.Y.: Doubleday, Doran, 1945. 666 pp. One hundred selections from all fields of science from earliest times, with introductory notes. Selected readings, pp. 641-643, contains 52 books, anthologies and periodicals. Some 10 selections concern pioneer works related to electricity and electronics from the mid-19th century.

205 Chase, Carl T., The Evolution of Modern Physics. New York: D. Van Nostrand, 1947. 204 pp. A popular history of ideas and experiments from ancient times. Includes electrons, electron waves, photoelectricity and related topics.

206 Taylor, F. Sherwood, A Short History of Science and Scientific Thought. New York: W.W. Norton, 1949. 368 pp., 53 illus., 28 plates. This expanded version of Science Past and Present (203), "With readings from the

great scientists from the Babylonians to Einstein," has additional material on atoms, particles and waves.

207 Dampier, William C., A History of Science and its Relations with Philosophy and Religion. Cambridge: University Press, 1949. 527 pp., numerous footnote refs. A standard work with minor technical interest.

208 Chalmers, Thomas W., Historic Researches. Chapters in the History of Physical and Chemical Discovery. London: Morgan, 1949; New York: Charles Scribner's Sons, 1952. Reprinted 1968. 288 pp., 85 illus. Biographical Notes, pp. 259-274, has 34 entries. This useful survey is limited by the absence of documentation. Originally issued in 38 articles in The Engineer, 1944-1948.

209 Wightman, William P.D., The Growth of Scientific Ideas. New Haven: Yale University Press, 1951, 1953. 495 pp., 35 illus., 8 plates. A "guide to the study of the development of scientific thought." In two parts: matter and motion, nature and life. Includes electricity, magnetism, particles and fields. A few items are listed as sources at the end of each chapter. A short, general bibliography lists 17 items, and there is a condensed chronology up to 1895.

210 Taylor, F. Sherwood, An Illustrated History of Science. London: William Heinemann; New York: Frederick A. Praeger, 1955. 178 pp., 120 photos and diags., including 59 drawings by A. R. Thomson. Notes on these drawings, pp. 168-171. Various early electrical experiments and discoveries up to the mid-19th century are discussed and portrayed; H. Hertz, W. Crookes, J. J. Thomson and their work are also mentioned. References not given.

211 Schwartz, George, and Philip W. Bishop (eds.), Moments of Discovery. The Origins of Science. New York: Basic Books, 1958. 2 vols., pp. 497, 503-1005. Index in Vol. 2. Extracts from scientific writings with introductions. Chap. 8, pp. 843-950, on electrons, atoms, and rays, is of present interest. Some illus., no documentation. See 219.

212 Shamos, Morris H., Great Experiments in Physics. New York: Henry Holt, 1959. 370 pp., some line drawings. Each chapter consists of a biographical introduction with references and one or more extracts from the origi-

nal literature. Numerous marginal notes and some footnotes, mainly explanatory. Reading list with each chap. Parts of present interest are: Electromagnetic waves, H. Hertz, pp. 184-197; X-Rays, W. K. Röntgen, pp. 198-209; The electron, J. J. Thomson, pp. 216-231; The photoelectric effect, A. Einstein, pp. 232-237; The elementary electric charge, R. A. Millikan, pp. 238-249; The electromagnetic field, J. C. Maxwell, pp. 283-300.

213 Nobel Lectures, Physics 1901-1921; also 1922-1941, 1942-1962. Amsterdam, London, New York: Elsevier, 1964-1967. Presentation speeches and laureates' biographies. The following are of particular interest. 1901, Wilhelm C. Röntgen. 1902, Hendrik A. Lorentz and Pieter Zeeman. 1905, Philipp Lenard. 1906, Joseph J. Thomson. 1909, Guglielmo Marconi and Carl F. Braun. 1921, Albert Einstein. 1923, Robert A. Millikan. 1928, Owen W. Richardson. 1937, Clinton J. Davisson and George P. Thomson. 1939, Ernest O. Lawrence. 1947, Sir Edward V. Appleton. 1956, William Shockley, John Bardeen and Walter H. Brattain. Some lectures are illustrated with photos and diags., and some have lists of publications, patents or other refs. Valuable source material. See also 372.

214 Siegbahn, Manne, "The Physics Prize." Nobel. The Man and His Prizes, 1962 (506), pp. 441-518. Commentary on the state of physics in 1901 and its progress to recent years, with brief descriptions of the physicists' work, discoveries, and awards. Names of particular interest include W. C. Röntgen, H. A. Lorentz, P. Lenard, J. J. Thomson, G. Marconi, C. F. Braun, R. A. Millikan, O. W. Richardson, E. V. Appleton, W. Shockley, J. Bardeen, W. H. Brattain.

215 Davis, Watson, The Century of Science. New York: Duell, Sloan and Pearce, 1963. 313 pp., 224 photos, ports., diags., and charts. This broad survey from the mid-1890s glances at electrical communications, pp. 72-83, and electronics and automation, pp. 150-169. No refs.

216 Forbes, Robert J., and E. J. Dijksterhuis, A History of Science and Technology. Harmondsworth, Middlesex; Baltimore: Penguin Books, 1963, 2 vols. Vol. 1, Ancient times to the seventeenth century. 294 pp., 8 plates, 20 diags. Bibliography, pp. 261-274, 342 items in 16 groups, mainly by chapters. Vol. 2, The eighteenth and nineteenth centuries, pp. 305-536, 8 plates, 6 diags. Bibliography,

pp. 511-516, 135 items in 11 groups by chapters. Chaps. 23, 24, pp. 431-472: Magnetism and electricity in the eighteenth and nineteenth centuries; Steel and electricity change the world. Very slight treatment of communications and electronics.

217 Reichen, Charles-Albert, A History of Physics. New York: Hawthorne Books, 1963. The New Illustrated Library of Science and Invention, Vol. 8. 111 pp., over 140 illus., some full page and some in color. Illustrated chronology, pp. 106-110, from 3000 B.C. to 1962. A highly illustrated picture essay with brief supporting text. Lively, popular treatment for the general reader. Useful list of picture sources, p. 111. Slight technical interest. No refs., no index.

218 Wright, Stephen (comp.), Classical Scientific Papers: Physics. London: Mills and Boon, 1964. 393 pp., plates, diags., charts, tables. Facsimile reproductions of famous papers, with an introduction by the compiler. There are 21 selections arranged in 4 groups: Radioactivity, The atom, Further developments, Some tools of the trade. Several items are of electronics interest.

219 Hurd, D. L., and J. J. Kipling (eds.), The Origins and Growth of Physical Science. Harmondsworth, Middlesex; Baltimore: Penguin Books, 1964, 2 vols. Vol. 1, 343 pp., 8 plates, other text illus. Vol. 2, 427 pp., a few illus. in the text. Based on Moments of Discovery (211), contains extracts from scientific writings with introductions. Vol. 2, Pt. 3, The electromagnetic synthesis, pp. 149-266, from Newton to Hertz. Pt. 5, The breakdown of classical physics, pp. 319-409, includes X rays, the electron, and quantum physics.

220 Shepherd, Walter, Outline History of Science. New York: Philosophical Library, 1968. 142 pp. Chronological listing of notable events from 5000 B.C., divided into seven periods. Alphabetical list of notable inventions and discoveries, pp. 97-110. Laws and principles, pp. 111-128, 133 entries from aberration of light to Zeeman effect, with brief descriptions. List of outstanding scientists since the Renaissance, pp. 129-134, arranged alphabetically. Bibliography, pp. 135, 136, 82 items.

221 Taton, René (ed., A. J. Pomerans, trans.), History of Science. London: Thames and Hudson; New York:

Basic Books, 1963-1966, 4 vols. Vols. 1, 2, Ancient and medieval to 1800. Vol. 3, Science in the nineteenth century, 623 pp., 48 plates, 20 text illus. In 6 pts., mathematics, mechanics and astronomy, physical science, geological sciences, biological sciences, science and society. Bibliography, pp. 529-547 (electricity and magnetism, p. 537, 14 entries), and separate lists for Pt. 6. Vol. 4, Science in the twentieth century, 638 pp., illus. See 244.

B. Engineering and Technology

222 Singer, Charles, and others (eds.), A History of Technology. London: Oxford University Press, 1954-1958, 5 vols. Vol. 4, The industrial revolution, c1750 to c1850, 728 pp., illus. Chap. 22, Telegraphy. Vol. 5, The late nineteenth century, c1850 to c1900, 888 pp., illus. Chaps. 9, 10, The rise of the electrical industry, see 335. All vols. are highly illus., with text diags. and full-page plates. There are 3 indexes in each vol; personal names (with dates and descriptions), place names, subjects. This monumental work treats a wide range of subjects in great detail, from earliest times to 1900. There is considerable emphasis on European, American and British developments, especially the latter, in the last two volumes. Many details of design, construction, techniques and applications are given, supported with excellent illustrations. The text figures and plates are listed separately, with sources. Almost all chapters have a list of numbered references and a bibliography.

223 Fleming, Arthur P. M., and Harold J. Brocklehurst, A History of Engineering. London: A. & C. Black, 1925. 312 pp., 1 plate. About 235 footnotes. Bibliography, pp. 299-304, more than 200 entries (without dates and publishers) arranged by chapters. Chap. 10, The age of electricity, pp. 234-272, includes a very light treatment of telegraphs, telephones, and radio communication.

224 Lilley, Samuel, Men, Machines and History. The Story of Tools and Machines in Relation to Social Progress. London: Cobbett Press, 1948. Revised ed., New York: International Publishers, 1966. 352 pp., 48 plates, 23 text illus. A survey from 3000 B. C. There are brief treatments of the telegraph, telephone, radio, electronic industries, radar, electron tubes and transistors. Electronic computers and applications, pp. 231-251, and automation, pp. 252-276, are dealt with more extensively. Some

of the 200 footnotes include refs. Emphasizes political, military and national policies, the characteristics of technological progress, society and technology, and national achievements, particularly in regard to U.K., U.S.A., and U.S.S.R. The restrictive practices of capitalistic forces, especially big business and government, and their effects on technological progress are discussed in detail.

225 Forbes, Robert J., Man the Maker. A History of Technology and Engineering. New York: Henry Schuman, 1950. 355 pp., 18 plates, 27 illus. in the text. Bibliography, pp. 331-337, contains about 90 items grouped by chapters. Chap. 10, pp. 297-306, Steel and electricity, makes only passing references to the telegraph, telephone, radio and television. 2nd ed., 1958. 381 pp., 41 plates and text illus.

226 Dunsheath, Percy (ed.), A Century of Technology: 1851-1951. London: Hutchinson's Scientific and Technical Publications, 1951. 346 pp., 10 plates, 5 diags. This overview of progress has two short chapters of present interest: Electrical engineering, pp. 160-173; Telecommunications, pp. 269-283. For general readers. There are no references.

227 Kirby, Richard S., and others, Engineering in History. New York: McGraw-Hill, 1956. 530 pp., 175 illus. General bibliography, pp. 515-516, 17 items; also short chap. bibliographies and numerous footnotes. Chap. 11, Electrical engineering, pp. 327-373, includes a brief survey of telecommunications, pp. 336-351.

228 Oliver, John W., History of American Technology. New York: Ronald Press, 1956. 676 pp., over 300 numbered chap. notes with about as many other chap. refs. Telecommunications dealt with very briefly; pp. 433-440, telegraph, ocean cable, telephone, wireless; pp. 496-502, radio from 1900 to 1915; pp. 540-547, radio broadcasting, telephone, television, radar.

229 Klemm, Friedrich (Dorothea W. Singer, trans.), A History of Western Technology. New York: Charles Scribner's Sons, 1959. 401 pp., 24 plates, 59 illus. 50 text footnotes, pp. 381, 382. Refs. to sources, pp. 383-388, 219 entries. Select bibliography, pp. 389-392, has 100 entries. Selections from contemporary literature with introductory and co-ordinating comments. Very little of electrical interest.

230 Gartmann, Heinz (Alan G. Readett, trans.), Science as History. The Story of Man's Technological Progress from Steam Engine to Satellite. London: Hodder and Stoughton, 1960. 348 pp., 32 plates, 74 other illus. Bibliography, pp. 337-343, includes many German items. More on technology than science. Chap. 4, pp. 122-181, radio, telephone, radar and television. Chap. 6, pp. 229-260, automation and computers. Index gives names only. No refs.

231 Armytage, W. H. G., A Social History of Engineering. London: Faber and Faber, 1961; Cambridge, Mass.: M.I.T. Press, 1961, rep. 1966. 378 pp., 32 plates. List of illus. and sources, pp. 13-16. Bibliography, pp. 335-353, contains about 350 entries related to individual chapters and to general works. List of professional institutions in Britain, pp. 354-357. Meager treatment of electricity, electronics and telecommunications.

232 Kranzberg, Melvin, and Carroll W. Pursell (eds.), Technology in Western Civilization. New York: Oxford University Press, 1967, 2 vols. Vol. 1, The emergence of modern industrial society, earliest times to 1900. 802 pp., diags., photos, maps, tables, readings and refs., pp. 745-774. Vol. 2, Technology in the twentieth century. 772 pp., diags., photos, charts, tables. Readings and refs., pp. 709-739. There is a list of contributors in both vols. Relevant chapters on electrical communications and the computer are listed separately, see 309, 1104. A comprehensive work by specialists.

233 Calder, Ritchie. The Evolution of the Machine. New York: American Heritage, 1968. First vol. in a new series of The Smithsonian Library. 160 pp., about 150 illus., some full-page and some in color. Illus. chronology, pp. 129-138, from earliest times to 1967. Illus. of patents, pp. 139-145. Illus. biographies, pp. 146-153. Reading list, pp. 154, 155, has 62 entries. List of picture sources, p. 160. A bright pictorial essay with brief textual sketches intended for the layman. No refs. Very little electrical or electronic content.

234 Fyrth, Hubert J., and Maurice Goldsmith, Science, History and Technology. London: Cassell; Book 1, 1965, Book 2 (3 Pts.), 1969. 260, 89, 138, 107 pp. Each book deals with perioas: 800-1840s, 1840s-1880s, 1880s-1940s, 1940s-1960s. The first book contains 16 plates, the others have 4 plates. Each book contains a date chart,

chap. notes, classified reading list and index with biographical dates. Much factual material--names, dates, events--with major attention on the social and political scenes. Numerous quotations and extracts, but little documentation. Comprehensive survey of the title subjects with very light treatment of electricity and electronics.

235 Daumas, Maurice (Eileen B. Hennessy, trans.), A History of Technology and Invention. New York: Crown, 1969, 2 vols. Vol. 1, Origins of technical civilization, from the beginnings to the Middle Ages, 596 pp., 113 illus., 66 plates. Vol. 2, First stages of mechanization, from the mid-15th century to the middle of the 18th century. 694 pp., 317 illus., 53 plates. Vol. 1, pp. 572-576, General bibliography, 112 items in 16 groups. There are bibliographies for most chapters in both vols. This encyclopedic work, originally published in France in 1962, covers all areas of technology. Two further vols. (3, 1700-1850; 4, 1850-) will bring the record up to the present.

C. Electricity

236 Mottelay, Paul F., Bibliographical History of Electricity and Magnetism, Chronologically Arranged. London: Charles Griffin, 1922. 673 pp., 14 plates. A comprehensive record with details of publications in all languages from earliest times to 1821. Entries contain particulars of inventions, discoveries, experiments, demonstrations and theories, with names and dates. A list of references and sources, frequently quite extensive, is given with almost all entries. Additional material is contained in five appendixes, pp. 501-564. These include an account of early writers, discoveries made by William Gilbert, a listing of the Philosophical Transactions of the Royal Society, list of additional works, and Mercator's Projection. The index, pp. 565-673, is exceptionally detailed with numerous cross references and additional data.

237 Benjamin, Park, The Age of Electricity. From Amber-Soul to Telephone. New York: Charles Scribner's Sons; London: Cassell, 1886, 1901. 381 pp., 4 plates, 143 illus., diags. Historical highlights with simple descriptions and extracts.

238 Walsmley, Robert M. (rev.), Electricity in the Service of Man. London: Cassell, 1890. (German origi-

nal by Alfred R. von Urbanitzky, Vienna: 1885, 1092 pp.) R. Wormall (ed.), with introduction by John Perry. 891 pp., nearly 870 illus. "A popular and practical treatise on the applications of electricity in modern life." Principles of electrical science, pp. 1-224. Technology of electricity, pp. 227-470. Practical applications, pp. 471-876. Telephone, pp. 691-773; photophone, pp. 774-785; telegraph, pp. 786-876. Historical introduction to most sections. A comprehensive work with many excellent engrs. Also 1896, 976 pp. A revised and expanded ed. was published in 3 vols. Vol. 1, The history and principles of electrical science, 1911, 840 pp. Vol. 2, Pt. 1, The technology of electricity, 1913, 692 pp., Pt. 2, 1919, 724 pp.

239 Electricity in Daily Life. New York: Charles Scribner's Sons, 1890. 288 pp., 125 illus., including 6 plates. This popular account by 10 contributors consists of articles that originally appeared in Scribner's Magazine. "The Telegraph of Today," by Charles L. Buckingham, pp. 140-172. There are other chapters on telegraph cables, electricity in naval and land warfare, electricity in the household, and electricity in relation to the human body, in addition to parts on motors, lighting, and the electric railway. The historical material is supplemented by engravings.

240 Tunzelman, G.W. de., Electricity in Modern Life. New York: P.F. Collier, 1900. 285 pp., 93 illus. including 11 plates. A survey for the general reader up to 1889, with historical comments.

241 Munro, John, The Romance of Electricity. London: Religious Tract Society, 1893. 320 pp., 57 illus. Natural electricity occupies about 200 pages, with numerous accounts of lightning, fireballs, St. Elmo's fire, aurorae and electrical discharges in air and in vacua. The last part of the book treats curiosities of the telegraph, telephone, and electric light.

242 Houston, Edwin J., Electricity One Hundred Years Ago and To-day. New York: W.J. Johnston, 1894. 199 pp. The text of a lecture with additional copious notes and extracts from early works. About half of the book embraces history before 1800. Primarily on basic discoveries and inventions, with mention of the telegraph and the telephone. Many refs.

243 Mendenhall, Thomas C., A Century of Electricity.

Boston; New York: Houghton, Mifflin, 1887. 229 pp., 28 engrs. Chaps. 4, 5, pp. 88-110; 111-150, on the telegraph and submarine cables. Chap. 9, pp. 202-212, on the telephone. Historical sketch for the layman. No refs.

244 Bauer, E., "Electricity and Magnetism (1790-1895)." History of Science, Vol. 3, Chap. 4, pp. 178-234. Includes Maxwell's theory, the work of Hertz, conduction in rarefied gases and the beginnings of the electron theory. A survey from the scientific viewpoint. Not documented. See 221.

245 Jerrold, Walter, Electricians and Their Marvels. New York: F. H. Revell, 1896.

246 Munro, John, The Story of Electricity. London: George Newnes; New York: D. Appleton, 1896, 1915. 225 pp., 103 illus. A popular treatment with a light historical touch.

247 Houston, Edwin J., Electricity in Every-Day Life. New York: P. F. Collier, 1904. 3 vols. Vol. 1, 584 pp., 19 plates (3 in color), 214 illus. The generation of electricity and magnetism; electric machines, high-voltage electricity, gaseous discharges, X rays, magnetism, electric cells, other electric sources, electromagnetic induction. Vol. 2, 566 pp., 19 plates (3 in color), 307 illus. The electric arts and sciences; dynamos, alternators, electric lamps and lighting, electric power, motors, power transmission, electric locomotives and railways. Vol. 3, 609 pp., 18 plates (2 in color), 270 illus. The electric arts and sciences; electrochemistry, telephony, telegraphy, annunciators and alarms, electric heating, electro-therapeutics. The phonograph, microphone and photophone, pp. 186-201. Facsimile and time telegraphy, pp. 366-384. Wireless or space telegraphy, pp. 393-426. A thorough survey with historical background.

248 Durgin, William A., Electricity, Its History and Development. Chicago: A. C. McClurg, 1912. 176 pp., 28 plates. A popular survey for the layman.

249 Trowbridge, John, The Advance of Electricity Since the Time of Franklin. Cambridge, Mass.: Harvard University Press, 1922. 183 pp.

250 Sharlin, Harold I., The Making of the Electrical Age. New York: Abelard-Schuman, 1963. 248 pp., 28

diags. and photos. Text refs. listed in Notes, pp. 231-242, 316 items. A combined survey of communications and power. Most of the book is concerned with developments up to 1900. Chap. 3, pp. 72-95, treats Maxwell's theory, Hertz's experiments and the early work of Marconi. Chap. 4, pp. 96-130, deals briefly with commercial radio up to the early 1920s. Computers and automation are scanned in the final chapter, pp. 218-230. A condensed story for the nonspecialist.

251 Greenwood, Ernest, Amber to Amperes, the Story of Electricity. New York: Harper and Brothers, 1931. 332 pp., 14 illus., mainly plates. Short bibliography, pp. 319-321, 51 entries without dates or other details. The telegraph, telephone and radio are treated in pp. 162-223. A general, nontechnical history with some extracts and sources, otherwise not documented.

252 Miller, Dayton C., Sparks, Lightning, Cosmic Rays: An Anecdotal History of Electricity. New York: Macmillan, 1939. 192 pp., 90 illus. 155 footnote refs., mainly to original papers and articles, some multiple entries. A record of "Christmas Week" lectures for young people given at the Franklin Institute, Dec. 1937. The first two lectures, pp. 3-103, are on early electricity, high-voltage electrostatic generators, Benjamin Franklin and his experiments. The third lecture, pp. 107-185, deals with radio, electricity in gases, cathode rays, X rays, and cosmic rays. A readable survey.

253 Still, Alfred, Soul of Amber. The Background of Electrical Science. New York: Murray Hill Books, 1944. 274 pp. History from earliest times, development of ideas, electrical machines, apparatus and instruments, batteries, the telegraph, dynamos, and theories of modern times. Chap. refs., 213 entries. Well documented. Chronological table, pp. 260, 261, 33 entries from 1600.

254 Zeluff, Vin, and John Marcus, "100 Years of Electricity and Electronics." Sci. Am., 173:7-12, 14, 16, July 1945. 13 photos. Power developments, electric lighting, X rays, wire communications, submarine cables, wireless and radio, electron tubes, television, radar, industrial electronics.

255 Still, Alfred, Soul of Lodestone. The Background of Magnetical Science. New York: Murray Hill Books, 1946. 233 pp. History since ancient times through the

work of William Gilbert, the magnetic compass and theories of magnetism. Chap. refs., 153 entries. Well documented.

256 Mann, Martin, Revolution in Electricity. London: John Murray, 1962. 171 pp., 138 illus. An informal account of electricity, with emphasis on transistors, solar cells, supermagnets and electronic light. Some historical background. No refs.

257 Canby, Edward T., A History of Electricity. New York: Hawthorne Books, 1963. The New Illustrated Library of Science and Invention, Vol. 6. 111 pp., more than 140 illus., some full-page and some in color. Illus. chronology, pp. 106-110, from 600 B. C. to 1962. A handsome picture essay with brief supporting text. List of picture sources, no refs.

258 Soanes, Sidney V., "A Philatelic History of Electrical Science." IEEE Spectrum, 7:80-85, Jan. 1970. 4 photos of 71 stamps. Table of stamps commemorating space flight. From Otto von Guericke (1680) to modern times, including about 30 events related to telecommunications.

D. Electrical Engineering

259 Jarvis, C. MacKechnie, "The History of Electrical Engineering." J. IEE., 1(NS):13-19, 145-152, 280-286, 566-574, Jan., Mar., May, Sept., 1955. 2:130-137, 584-592, Mar., Oct., 1956. 3:310-319, June, 1957. 4:298-304, June, 1958. 67 illus., 52 refs. Electrical science, electric light, rotating machines, telegraph, central stations. A survey from the 13th century to the rise of the power industry in the 1880s.

260 "Our 2000th Issue." Electrician, 77:789-896, Sept. 15, 1916. 24 historical articles covering 40 years (pp. 793-877). 4 are listed elsewhere, see 1270, 1533, 1550, 1582.

261 Fleming, John A., Fifty Years of Electricity. The Memories of an Electrical Engineer. London: Wireless Press, 1921. 371 pp., port., 110 plates, 167 diags. Introduction, pp. 1-49, gives an overview of 50 years from 1820. Chap. 1, pp. 50-108, Telegraphs and telephones from 1870 to 1920. Chaps. 2-5, pp. 109-265, Power, lighting, electrical machines and applications. Chap. 6, pp. 266-306,

Electric theory and measurements, includes theories and experiments related to electromagnetic waves, the discharge of electricity through rarefied gases, X rays, and the discovery of the electron. Chap. 7, pp. 307-345, Wireless telegraphy and telephony. App., pp. 353-361, 128 entries of the author's works and lectures. A valuable account by a foremost engineer and educator. Some names, dates and refs. in the text, occasional footnotes, otherwise not documented.

262 Howe, G. W. O., "A Hundred Years of Electrical Engineering." Brit. Assoc. Adv. Sci., 1924, pp. 178-189; Engineer, 138:166-168, Aug. 8, 1924; Sci. Mon., 19:290-304, Sept. 1924; Nature, 114:277-280, 1925. Mainly power engineering, with brief mention of the telegraph, telephone and radio.

263 "Fiftieth Anniversary Number." Elect. Eng., 53:641-848, May 1934. A valuable historical collection of 51 articles and photo groups related to the early days and subsequent developments of electrical engineering and the American Institute of Electrical Engineers. Items about the Institute include the 50th anniversary, the first half century, organization, work on standardization, early headquarters, and some major events since 1884. The biographical section treats 92 subjects, most with ports. There are 23 photos of telegraph, telephone and radio equipment. A chronology of major events occupies two pages. Most of the issue is devoted to power engineering and heavy machinery. Some articles are listed separately, see 264, 265, 286, 475, 476.

264 Jackson, Dugald C., "The Evolution of Electrical Engineering." Elect. Eng., 53:770-776, May 1934. Reminiscences concerning the International Exhibition of 1884, the state of electrical science at that time, electrical engineering education, training, schools and curricula.

265 Thomson, Elihu, "Some Highlights of Electrical History." Elect. Eng., 53:758-767, May 1934. 7 photos, port. The author recalls a number of events in which he played leading roles. Early experiments and lectures, development of electric lighting, power, apparatus and instruments. Problems of design, development of electrical communications, early experiments in electronics.

266 Cope, H. W., "Half Century of Engineering Progress." Electric J., 33:3-16, Jan. 1936. Story of George Westinghouse, the company, and the electrical industry.

267 National Electrical Manufacturers Association, A Chronological History of Electrical Development: From 600 B.C. New York: The Association, 1946. 106 pp., App., 23 pp., Index, 13 pp. Brief listings, almost wholly on electrical power. Member companies with historical data (original name, founding date, name of founder and title, first president) are listed in the App., 462 entries.

268 Ferguson, J.D., "A Synoptic Review of Electrical Engineering Progress, Particularly during the Last Quarter Century." J. IEE., 94 (Pt.1):73-81, Feb. 1947. Fundamentals, telegraphy, telephony, radio communication, electronics, as well as power engineering. No refs.

269 Everett, S. Lee, "Light's Diamond Jubilee and the Engineer." Elect. Eng., 73:1057-1062, Dec. 1954. A survey of progress in electrical engineering with reference to recipients of the Edison Medal. Highlights of Edison's inventions, 18 entries from 1868. Names and dates of award winners, 43 entries from 1903. Other items relating to light and power, the telephone and electrical communication, radio and television, educators, electrical manufacturing and research.

270 "75th Anniversary Issue." Elect. Eng., 78:413-614, May 1959. 22 articles in 4 groups; basic sciences, communications, industry and applications, electric power. Some of these are entered separately. See 271, 299, 300.

271 Suits, C. Guy, "Basic and Applied Engineering Research." Elect. Eng., 78:415-422, May 1959. 15 photos. On electrical developments since 1934.

272 Westcott, G.F. (comp.), H.P. Spratt (rev.), Synopsis of Historical Events: Mechanical and Electrical Engineering. London: H.M.S.O., 1960. 44 pp. A listing of events from c.500,000 B.C. to 1958. Each entry gives name(s), brief description, and source refs. Some items mention related events, dates, etc., and some include a variety of supplemental material, such as construction details, or technical data, and applications, with comments. Only a few entries concern communications or electronics. References, pp. 41-44, give author, title, publication date.

273 Dunsheath, Percy, A History of Electrical Engineering. London: Faber and Faber, 1962. 368 pp., 48 plates, 71 diags. 253 chap. refs. Chronological table, pp.

347-357, from 1269 to 1961. A basic and authoritative treatment with emphasis upon power applications, wire services and general engineering, particularly from the British viewpoint. Electronics is treated superficially in Chap. 17, pp. 266-289, The electron in engineering. Other parts likely to prove useful include Science in telecommunications, pp. 244-253, and Professional organization, pp. 319-332. The final chapter is a brief attempt to place the progress of electrical science and engineering against a social and historical background.

274 Kingsford, P. W., Electrical Engineering: A History of the Men and the Ideas. London: Edward Arnold; New York: St. Martin's Press, 1970. 268 pp., 20 illus. on 12 plates. The contributions of some two dozen pioneers highlight this story of progress, mainly from 1800. There are brief biographical details and numerous extracts from books, correspondence and other documents. Three chaps. deal with radio, television and broadcasting, and radar. Over 40 pp. are concerned with electrical workers and trade unions from the 1860s to 1960. A few books are listed with most chaps. Refs. not given.

E. Telecommunications

275 Marland, Edward A., Early Electrical Communication. New York: Abelard-Schuman, 1964. 220 pp., 8 plates, 35 diags. Fully documented with 350 footnote refs. Chronological table of contributors to the telegraph, pp. 201, 202. Biographical memoranda, pp. 203-208, has 30 brief but useful entries. Bibliography, pp. 209-212, lists 62 items. A list of principal journals referred to in the text, with page numbers, is given in pp. 213, 214. A survey of telecommunications during the 19th century, with emphasis upon inventions and apparatus.

276 McNicol, Donald (comp.), "A Chronological History of Electrical Communication--Telegraph, Telephone and Radio." Radio Engineering, Vols. 12-14, Jan. 1932 to Aug. 1934, one page per month. A record of discoveries, inventions, innovations and patents, with names, dates, necrology and statistics, from 2000 B.C. to 1910. There are 32 parts with 1276 numbered entries. Many items concern power engineering, communication companies, manufacturers, and company statistics, with emphasis on American personalities and affairs.

277 "Communicating Over Great Distances." Sci. Am., 112:531, 532, 566, 570-574, June 5, 1915. Illus. This "record of achievement from Morse to Marconi" is a compact survey of progress of the telegraph, telephone and wireless. Facsimile telegraphy and phototelegraphy, pp. 571-574.

278 Towers, Walter K., Masters of Space. New York: Harper and Brothers, 1917. 300 pp., plates, ports., numerous illus. Morse and the telegraph, Thompson and the cable, Bell and the telephone, Marconi and the wireless telegraph, Carty and the wireless telephone. 2nd ed., From Beacon Fire to Radio; the Story of Long-Distance Communication, 1924.

279 Pupin, Michael I., "Fifty Years' Progress in Electrical Communications." Science, 64:631-638, Dec. 31, 1926. Also, J. AIEE., 46:59-61, 171-174, Jan., Feb. 1927. General survey. Address to the American Association for the Advancement of Science, Dec. 27, 1926.

280 Brown, Frank J., The Cable and Wireless Communications of the World. London: Pitman, 1927. 148 pp. 2nd ed., 1930, 153 pp. See also 288.

281 Woodbury, David O., Communication. New York: Dodd, Mead, 1931. 280 pp., 14 plates. Bibliography, pp. 273-278, 75 items, incomplete and partly repetitious. A popular story of the telegraph, telephone and radio.

282 Crawley, Chetwode G., "The Story of Electrical Communications." Television, 1:35-37; 34, 35; 2:7, 8, 73-75, 125-127, 179-181, 225, 227, 290-292, 332-334, 390-392, 398, 439-441, 489-492, 539-541, 605, 606; 3:59-61, 134, 135, 146, 237-239. 1928-1931. See next entry.

283 Crawley, Chetwode G., From Telegraphy to Television. The Story of Electrical Communication. London: Frederick Warne, 1931. 212 pp., 24 plates, 6 text illus. A survey for the general reader.

284 Clark, George H., The Calendar of Wireless-Radio-Television. Washington: Smithsonian Institution. Clark Radio Collection, typescript, 14 pp., n.d. Brief entries from 640 B.C. to 1932. Basic material as in Dunlap's Radio and Television Almanac (295).

285 Squier, George O., Telling the World. Baltimore:

Williams and Wilkins, 1933. 163 pp., port., 1 plate. Chap. refs., pp. 156-162, contain 122 detailed entries of books, articles and patents. Readable story of electricity and magnetism, telegraph, telephone, radio. Chap. 5, pp. 124-155, is about the U. S. Signal Corps and wired radio.

286 Gherardi, Bancroft, "Communication--Past and Present." Elect. Eng., 53:745-752, May 1934. Illus. Brief survey from 1884. 97 refs. to AIEE publications.

287 Harlow, Alvin F., Old Wires and New Waves; the History of the Telegraph, Telephone, and Wireless. New York: Appleton-Century, 1936. 548 pp., 72 plates and illus. Extensive bibliography, pp. 527-538, has 187 detailed refs. A broad survey, not documented.

288 The Cable and Wireless Communications of the World. Cambridge: Cable and Wireless, 1939. 282 pp. A collection of lectures and papers, 1924-1939. See also 280.

289 Kohlhaas, H. T., "Milestones of Communication Progress." Elect. Comm., 20:143-185, 1942. 31 illus., 159 refs., almost all concerning articles in previous issues of this journal. On telephone, telegraph and radio equipment and installations, including cables, insulation, and early microwave trials. An international survey of progress over 20 years related to the member companies of IT&T.

290 Espenschied, Lloyd, "Electric Communications, the Past and Present Illuminate the Future." Proc. IRE., 31:395-402, Aug. 1943. Contains a time chart from 1840 to 1960, 2 refs. On information, frequency spectrum, and telecommunication services.

291 Mance, Harry O., International Telecommunications. London: Oxford University Press, 1943. 90 pp. One of 7 vols., International Transport and Communications.

292 Barrett, R. T., "A Century of Electrical Communication." Bell Tel. Mag., 23:43-51, Spring, 1944.

293 Coggeshall, Ivan S., "A Critique of Communication at the Centennial of the Telegraph." Proc. IRE., 32: 445-448, Aug. 1944. General historical remarks on the telegraph, its relationship to electrical engineering and world communications.

294 Still, Alfred, Communication Through the Ages. From Sign Language to Television. New York: Murray Hill Books, 1946. 201 pp., 24 photos and diags. 140 footnotes, including some refs. to sources. Chronological table, pp. 191-194, 87 entries from 1084 B.C. to 1936. A popular story with some philosophical features.

295 Dunlap, Orrin E., Radio and Television Almanac. New York: Harper and Brothers, 1951. 211 pp., 32 plates. Pages 194-198 list members of the Federal Radio Commission and the Federal Communication Commission; Presidents of the Institute of Radio Engineers, National Association of Broadcasters, Radio Manufacturers Association and Television Broadcasters Association. A popular outline history of radio and television based primarily on historic events and inventions. Entries give month and day, with brief description. Heavy emphasis on commercial and public events, especially American. A general reference without documentation. See 284.

296 Kelly, Mervin J., "Communications and Electronics." Elect. Eng., 71:965-969, Nov. 1952. Brief survey of highlights from the early 19th century.

297 Pratt, Haraden, "The First 50 Years of International Radio Communication." Communication and Electronics, No. 3, pp. 371-375, Nov. 1952. A brief survey.

298 "A Short History of Electrical Communications." Washington: FCC Bulletin No. 7, April 1953.

299 Engstrom, Elmer W., "The Transformation of Electronic Communications." Elect. Eng., 78:462-470, May 1959. 9 photos. Radio and television progress since the mid-1930s.

300 Green, E.I., "The Evolving Technology of Communication." Elect. Eng., 78:470-480, May 1959. 5 photos, 5 charts and diags. The telephone and transmission systems, growth of radio, electronic switching, data communication, waveguides, and global communication over the last quarter century. 3 refs.

301 Coggeshall, Ivan S., "The Compatible Technologies of Wire and Radio." Proc. IRE., 50:892-896, May 1962. An overview of relationships since 1913. 20 refs.

302 Dunlap, Orrin E., Communications in Space: From Wireless to Satellite Relay. New York: Harper and Brothers, 1962. 175 pp., 16 plates. A broad survey with a historical viewpoint, limited to American contributions. Intended for the general reader. Slightly documented; 87 footnotes provide supplemental data rather than specific refs.

303 Goldsmith, Alfred N. (ed.), "Fiftieth Anniversary Edition." Proc. IRE., Vol. 50, No. 5, pp. 529-1448, May 1962. In the section on "Communications and Electronics," pp. 657-1420, there are 113 papers in 28 groups, many of them on historical developments or with historical material. Several papers are cited elsewhere by subject.

304 O'Connell, James D., Alvin L. Pachynski, and Linwood S. Howeth, "A Summary of Military Communication in the United States--1860 to 1962." Proc. IRE., 50: 1241-1251, May 1962. 9 footnotes.

305 Howeth, Linwood S., History of Communications-Electronics in the United States Navy. Washington: G.P.O., 1963. 657 pp., 54 illus. Prepared under the auspices of Bureau of Ships and Office of Naval History. App. A, pp. 513-546, Chronology from 64 B.C. to 1945. Other Apps., pp. 547-610, on legislation, radio controls, frequencies, patents, equipment. Bibliography, pp. 611-613, has 80 items. This extensive work is in three parts: 1, The decade of development, pp. 1-202; 2, The golden age, pp. 207-394; 3, The electronic age, pp. 397-512. Exceedingly well documented with over 1500 footnotes. Liberal extracts and quotations from official records, documents, letters, memoranda and other archival sources pertaining to commercial organizations as well as government files.

306 Britain and Commonwealth Telecommunications. London: Central Office of Information, 1963. 36 pp., 5 tables. Reading list, p. 36, 14 items. Historical record of telegraph, telephone and cable systems, radio, television, organization and statistics. Prepared for British Information Service.

307 Fabre, Maurice, A History of Communications. New York: Hawthorne Books, 1963. The New Illustrated Library of Science and Invention, Vol. 9. 111 pp., over 140 illus., some full-page and some in color. Illus. chronology, pp. 106-110, from 20,000 B.C. to 1963. Picture essays with brief text. Popular treatment for the general reader. List of picture sources, p. 111. No refs., no index.

308 Michaelis, Anthony R., From Semaphore to Satellite. Geneva: International Telecommunications Union, 1965. 343 pp., 365 illus., many full-page, some colored. Pt. 1, The telegraph and the telephone, 1793-1932. Pt. 2, Radio, 1888-1947. Pt. 3, The Union after a century, 1947-1965. Conclusions. List of I.T.U. members, p. 337. Picture sources, pp. 339-342. List of reference books, p. 343, contains 48 titles. Highly illustrated, with brief supporting text. A short, accurate, but nontechnical account devoted mainly to international affairs. No index, no documentation.

309 Finn, Bernard S., "Electronic Communications." Technology in Western Civilization, Vol. 2, pp. 293-309. A compact survey of radio from the earliest days; Maxwell, Hertz, Marconi, transmitters and receivers before World War I, the crucial years 1912-1920, development of broadcasting, TV and FM, shortwave and microwave development, radar, progress during and after World War II. See 232.

310 Beck, Arnold H. W., Words and Waves. New York: McGraw-Hill, 1967. 255 pp., 20 photos, 58 diags. in color. Bibliography, pp. 245, 246, 19 books and articles. Chaps. 3 to 5, pp. 45-106, are on the history of the telegraph, telephone and radio. This "introduction to electrical communications" provides an excellent overview with some technical explanations designed for the layman. No refs.

311 Pierce, John R., The Beginnings of Satellite Communications. San Francisco: San Francisco Press, 1968. 61 pp., 8 illus. Refs., p. 60, 20 entries. App. 1, pp. 37-43, reprint of article by Arthur C. Clarke, "Extra-Terrestrial Relays," Wireless World, 51:303-308, Oct. 1945. App. 2, pp. 44-57, reprint of article by John R. Pierce, "Orbital Radio Relays," Jet Propulsion, 25:153-157, April 1955. Bibliography (by and about J. R. Pierce), pp. 58, 59, 26 entries; (of science fiction), p. 59, 13 items.

F. Inventors and Inventions

312 Routledge, Robert, Discoveries and Inventions of the Nineteenth Century. London: George Routledge, 1879, 594 pp. 9th ed., 1891. 681 pp., 432 illus. Electricity, pp. 387-443, includes machines, power and lighting. The electric telegraph, pp. 444-478; The telephone, pp. 478-489. Many good engrs.

313 Cockrane, Charles H., The Wonders of Modern Mechanism. London: J.B. Lippincott, 1896. 402 pp., 100 illus. Primarily on mechanical, physical and engineering science and inventions, there are short sections on electricity, its applications, and the telegraph. No index.

314 Byrn, Edward W., The Progress of Invention in the Nineteenth Century. New York: Munn, 1900; reprinted, Russell and Russell, 1970. 476 pp., 307 photos, diags., engrs. Chap. 2, pp. 7-14, Chronology of leading inventions of the nineteenth century, about 390 items listed by year. Chaps. 3, 4, pp. 15-37, The electric telegraph and the Atlantic cable. Chap. 8, pp. 76-87, The telephone. Chap. 22, pp. 273-283, The phonograph. Chap. 25, pp. 319-328, The Roentgen or X-rays. Wireless telegraphy is noted in pp. 26-30. "A cursory view of the century in the field of inventions." Many of the illustrations are reproduced from the Scientific American.

315 Cressy, Edward, Discoveries and Inventions of the Twentieth Century. London: George Routledge; New York: E.P. Dutton, 1914. 398 pp., 281 illus. 2nd ed., 1923. 458 pp., 342 illus., including 58 plates. Chap. 18, pp. 364-398, Wireless telegraphy and telephony. Chap. 20, pp. 422-453, Radium, electricity and matter. 3rd ed., 1930. 476 pp., 342 illus. 4th ed., rev., by James G. Crowther, 1955. 432 pp., 193 illus., including 64 plates. Radio and radar, pp. 346-382; The atom, pp. 401-428. 5th ed., 1966. 434 pp., 432 illus. Radio astronomy and radar, pp. 26-42. Computers, pp. 66-77. Radio and television, pp. 342-362. Rays and atoms, pp. 363-391. Names and dates in the text, but no refs. Each edition presents a nontechnical account of major advances and events in science, engineering and technology. Together they give an excellent panoramic view over 75 years.

316 Darrow, Floyd L., Masters of Science and Invention. New York: Harcourt, Brace, 1923. 350 pp., ports. London: Chapman and Hall, c.1936. 352 pp., 24 plates. This collection of brief biographies includes Faraday, Henry, Morse, Kelvin, Westinghouse, Edison, Bell, Marconi. Briefer biographies, pp. 326-343, 53 entries.

317 Kaempffert, Waldemar B. (ed.), Modern Wonder Workers: A Popular History of American Invention. New York: Charles Scribner's Sons, 1924. 577 pp., over 300 illus. The telegraph, telephone and radio are treated in pp. 286-378. No refs., no index.

318 "Calendar of Discovery and Invention." Nature, Vols. 119, 120, 1927. A single-column list in the weekly issues from Jan. to Dec. Miscellaneous topics, all fields of science and technology. About 350 entries by month, day and year, with brief descriptions.

319 Holland, Maurice, "Research, Science and Invention." A Century of Industrial Progress, pp. 312-334. This survey of industrial research mentions the telegraph, telephone, radio, and prominent U.S. companies. See 336.

320 Hawks, Ellison, The Book of Electrical Wonders. New York: Dial Press, 1935. 316 pp., 40 plates, 91 diags. On electricity, electric power, telephone, telegraph, radio, X rays, facsimile, television. A popular account without refs.

321 Abbot, Charles G., Great Inventions. Washington: Smithsonian Institution, 1932. Vol. 12, Smithsonian Scientific Series. 383 pp., 124 plates, 61 text illus. Chap. 3, pp. 49-71, Electrons and X rays. Chap. 4, pp. 72-112, Telegraphy and telephony. Chap. 5, pp. 113-134, Radio transmission. Some refs. are given in the text, and there are numerous extracts from original papers and patents.

322 Burlingame, Roger, Engines of Democracy. Inventions and Society in Mature America. New York: Charles Scribner's Sons, 1940. 606 pp., 33 plates, 90 other illus. Over 650 footnotes, mostly refs. Bibliography, pp. 545-568, 405 entries of books, articles, reports and papers. Bibliography--classified index, pp. 569-572, lists authors by subject. Events and inventions, a reference list, pp. 573-577, is a brief chronology from 1866 to 1929. A history of invention and society in the United States from about 1865. The telegraph, telephone, radio, television and kindred subjects are discussed, along with mention of English and European contributions.

323 Van Deusen, Edmund L., "The Inventor in Eclipse." Fortune, 50:132-135, 197, 198, 200, 202, Dec. 1954. The chief feature is a chronological chart of radio and television developments.

324 Wilson, Mitchell, American Science and Invention. A Pictorial History. New York: Simon and Schuster, 1954; Bonanza Books, 1960. 437 pp. Picture credits, pp. 428-430. Bibliography, pp. 431, 432, contains 247 entries of

articles and books, arranged alphabetically by author. This popularization is thoroughly illus. with about 1200 photos and engrs. The necessarily light treatment includes brief mention of the Atlantic cable, the war telegraph, Bell and the telephone, Edison and some of his inventions, Millikan and his work, de Forest and his early work in radio, and a quick look at electronics.

325 Tuska, Clarence D., Inventors and Inventions. New York: McGraw-Hill, 1957. 174 pp., illus. About imagination, inventing, inventions and patents.

326 Leithäuser, Joachim G. (Michael Bullock, trans.), Inventors of Our World. London: Weidenfeld and Nicolson, 1958. 257 pp., 16 plates and 12 other illus. A popular history of inventions and the creative process. Chap. 8, pp. 151-180, deals briefly with the telegraph, telephone, radio and television.

327 Jewkes, John, David Sawers, and Richard Stillerman, The Sources of Invention. London: Macmillan; New York: St. Martin's Press, 1958. 428 pp. Pt. 1, 260 pp., general survey of inventors, inventions, corporate research; 262 footnotes, source refs. and observations. Pt. 2, pp. 263-410, summaries of 50 case histories, with individual refs. and some footnotes. Nine cases refer to electrical and electronic inventions; including magnetic recording, radar, radio, television and the transistor.

328 Eco, Umberto, and G. B. Zorzoli (Anthony Laurence, trans.), A Pictorial History of Inventions. London: Weidenfeld and Nicolson, 1962. Also, The Picture History of Inventions. From Plough to Polaris. New York: Macmillan, 1963. 360 pp., lavishly illus. with 980 photos and diags., including 20 full-page color reproductions and others in color. Chap. 25, pp. 245-254, The telegraph and the telephone. Chap. 31, pp. 307-315, The electronic age. Chap. 33, pp. 327-338, The machine that thinks. Index of inventions, pp. 353-355. Index of illus., pp. 356-360, includes sources. A popular treatment without refs.

329 Larsen, Egon, A History of Invention. New York: Roy Publishers, 1962. 382 pp., 33 plates, 112 illus. Telegraph and telephone, pp. 256-276. Radio, pp. 277-290. Sound recording, pp. 291-296. Television, pp. 320-331. Everyday electronics, pp. 332-370 (radar, automatic controls, computers, electron microscope, radio astronomy). A popu-

lar survey of energy, transport and communications for the general reader. No refs. Also, Ideas and Invention. London: Spring Books, 1960. 384 pp., 41 plates.

330 Finn, Bernard S., "Controversy and the Growth of the Electrical Art." IEEE Spectrum, 3:52-56, Jan. 1966. 3 illus. Brief survey of some notable conflicts concerning inventions related to the telegraph, telephone, a-c power, and the rise of radio. 5 refs.

331 Carter, Ernest F., Dictionary of Inventions and Discoveries. London: Frederick Muller, 1966. 2nd ed., 1969. 204 pp. Entries are arranged alphabetically, with names, dates, brief descriptions, and cross references.

G. Industries

332 MacLaren, Malcolm, The Rise of the Electrical Industry During the Nineteenth Century. Princeton, N.J.: Princeton University Press, 1943. 225 pp., 154 illus. on 24 plates. 370 chap. refs. in bibliography, pp. 199-218. Chap. 2, pp. 32-64, Electric communication systems; early telegraphs, telephones and wireless experiments. A factual survey for the general reader.

333 Passer, Harold C., The Electrical Manufacturers 1875-1900. A Study in Competition, Entrepeneurship, Technical Change, and Economic Growth. Cambridge, Mass.: Harvard University Press; London: Oxford University Press, 1953. 412 pp., 3 diags., 11 other illus. Chap. notes, pp. 369-390, over 550 refs. Bibliography, pp. 391-400, 226 entries; company publications and records, books, articles and government documents. A thorough study of the title aspects concerned wholly with lighting, power and traction.

334 Bernal, J.D., Science and Industry in the Nineteenth Century. London: Routledge and Kegan Paul, 1953. 230 pp., 9 illus. Chap. 5, pp. 113-133, Electric light and power.

335 Jarvis, C. MacKechnie, "The Rise of the Electrical Industry." A History of Technology, Vol. 5, pp. 177-234. Chap. 9, The generation of electricity; Chap. 10, The distribution and utilization of electricity. 53 text illus. Bibliographies, 46 entries. The telegraph, telephone, and

wireless telegraphy (to 1900) are briefly mentioned. See 222.

336 Wile, Frederick W. (ed.), A Century of Industrial Progress. Garden City, N.Y.: Doubleday, Doran, 1928. 581 pp. 30 chaps. on developments in different industries and related technical and commercial activities. Three of these are listed elsewhere: Telegraphy and telephony (1623), Radio (1199), Research, science and invention (319). No index.

337 Glover, John G., and William B. Cornell (eds.), The Development of American Industries: Their Economic Significance. New York: Prentice-Hall, 1932. 932 pp., 96 photos, maps and charts. Chapters on 41 industries. 2nd ed., 1941. 3rd ed., 1951, 1121 pp., 78 photos, maps and charts. 4th ed., J.G. Glover and Rudolph Lagai, published by Simmons-Boardman, New York, 1959. 835 pp., 37 photos, maps and charts. Each chapter deals with general history, but emphasis is placed upon growth, organization and services of the respective contributors. Chaps. listed elsewhere: The telegraph industry (1536), and for the last edition, The electrical industry (342).

338 Maclaurin, William R., "The Role of the Large Electrical Firms in Wireless: 1912-1921." Invention and Innovation in the Radio Industry, Chap. 5, pp. 88-110. Also Chap. 7, pp. 132-152, "The Perennial Gale of Competition." Other parts also treat industrial aspects. See 1212.

339 Hall, Courtney R., History of American Industrial Science. New York: Library Publishers, 1954. 453 pp. Bibliographical note, pp. 419-423. Chap. 6, pp. 150-184, on the electrical and communications industries. Other scattered items of related interest. No refs.

340 Harris, William B., "The Electronic Business." Fortune, 55:136-143, 216, 218, 220, 224, 226; 134-138, 286-288, 290; 136-139, 292, 294, 296, 298, April, May, June, 1957. 23 charts, photos, diags. Business statistics, manufacturers, products and markets; components, electron tubes, transistors; computers, companies and business activities. A contemporary survey with some historical background.

341 Wilson, Thomas, "The Electronics Industry." The Structure of British Industry: A Symposium, Duncan Burn (ed.), 2 vols. Cambridge: University Press, 1958. Vol. 2, pp. 130-183. 7 tables, 2 charts. 58 footnotes, 9 items in a bibliography. The scope of the industry, innovation and growth, structure, productivity, competition and price-fixing, foreign trade, and the future.

342 [Westinghouse Electric Corporation]. "The Electrical Industry." The Development of American Industries, 4th ed., pp. 488-514. This retrospective view includes communications, electron tubes, solid-state devices, computers, and scientific research and development. Compare previous editions; see 337.

343 Kraus, Jerome, "The British Electron-Tube and Semiconductor Industry, 1935-62." Technology and Culture, 9:544-561, Oct. 1968. 54 footnote refs. The industry after World War I, the mid-1930s, television and radar, war production and postwar conversion, semiconductors, microelectronics. A study of trade factors and economic and technical developments.

See also 1212, Sects. 5C, 5D for company histories, and Sects. 13E, 13F for radio patents.

4. BIOGRAPHIES

Specialized biographical material and sources are grouped in this chapter; reference works, collections, and books and articles by or about individuals. The best sources for specific personal information are the professional journals listed in Sect. 2E (see also the indexes, catalogs and guides in Sects. 2C, 2D). These publications contain current notices related to group activities, membership, awards, obituaries and other personal items. Many serials include brief biographies of authors. Committee reports, convention records and news about organizations and industrial activities are other sources. Personal data can be found also in corporate publications, society directories, yearbooks, trade magazines and special annual reports. More detailed records of the life and work of prominent men are collected in the various biographical memoirs and obituaries published annually by professional societies. Many of the items listed in Chap. 3 also contain biographical material.

A. Biographical References

344 Poggendorff, Johann C., Biographisch-literarisches Handwörterbuch zur Geschichte der exacten Wissenschaften. Leipzig: J. A. Barth. 1863, 2 vols.; 1898, 2 vols.; 1904, 2 vols.; 1922, 1 vol.; 1931, 4 vols. In progress. This standard bio-bibliography has entries in English for the respective scientists and technologists. Author's works are listed.

345 [National Academy of Sciences]. Biographical Memoirs. Washington: The Academy, 1877, Vol. 1. Currently published by Columbia University Press. Contributed articles with portraits and bibliographies. Typically about 350 pp., 12 to 15 memoirs. Vol. 40, 1969, has a cumulative index to Vols. 1-40, pp. 349-354. Several memoirs are listed separately in Sect. 4C.

346 [Royal Society]. "Obituaries of Deceased Fellows chiefly for the period 1898-1904." Proc. Roy. Soc., Vol. 75, 1905. Also has a general index to obituaries in the Proceedings, Vols. 10-64, 1880-1899. Obituary notices from 1900 to 1932 were published in the Proceedings and in the Yearbook. See also 80, Vol. 3, 81, 349.

347 American Men of Science. The Physical and Biological Sciences. A Biographical Directory. New York: R.R. Bowker. 11th ed., 6 vols., 1965-67. About 130,000 brief biographies; includes engineers, teachers and others who have made notable contributions to applied science. Each entry gives details of education, vocation, specialty, affiliation and professional membership. There are 5 supplements of this current biography to date, covering the years 1966 to 1969. First ed., J. McKeen Cattell (ed.). New York: Science Press, 1906, 364 pp.

348 [Carnegie Library of Pittsburgh]. Men of Science and Industry; a Guide to the Biographies of Scientists, Engineers, Inventors, and Physicians, in the Carnegie Library of Pittsburgh. Pittsburgh: The Library, 1915. 189 pp. The titles indexed include collected works and publications of scientific and engineering societies.

349 [Royal Society]. Obituary Notices of Fellows of the Royal Society, 1932-1954. London: The Society, 1932-1954. 9 vols.

350 Biography Index. A Cumulative Index to Biographical Material in Books and Magazines. New York: H.W. Wilson. Vol. 1, Jan. 1946-July 1949, 1218 pp. About 40,000 entries with biographee's full name, date(s), nationality and profession, also portrait and obituary. The index to professions and occupations, pp. 1052-1218, lists biographees in about 1000 categories. These include electrical engineers, inventors, physicists, radio engineers, scientists, television engineers. Three-year cumulations to date.

351 Higgins, Thomas J., Biographies of Engineers and Scientists. Chicago: Illinois Institute of Technology, 1949. 62 pp.

352 [Royal Society]. Biographical Memoirs of Fellows of the Royal Society. London: The Society. Annual, Vol. 1, 1955 (current series). Signed articles with portrait and list of works.

353 Ireland, Norma O., Index to Scientists of the World from Ancient to Modern Times: Biographies and Portraits. Boston: F.W. Faxon, 1962. 43+662 pp. List of collections analyzed, with key to symbols, pp. xvii-xlii. Nearly 7500 scientists, inventors and engineers are listed alphabetically, with dates, brief identification and references. This broad coverage is based on 338 collections, ranging from specialized works and multivolume references to monographs, including juvenile and popular books. References to portrait sources is a special feature.

354 Higgins, Thomas J., "A Biographical Bibliography of Electrical Engineers and Electrophysicists." Technology and Culture, 2:18-32, 1961, Pt. 1, Books; 2:146-165, 1962, Pt. 2, Periodicals. About 1200 items arranged alphabetically by surnames. Recollections, surveys, evaluations, anniversary and obituary notices, reviews, and collections of works by or about individuals who made noteworthy contributions to electrical engineering.

355 Asimov, Isaac, Biographical Encyclopedia of Science and Technology. Garden City, N.Y.: Doubleday, 1964. 662 pp., 16 plates containing 79 ports. Brief biographies of more than 1000 scientists from the Greek era to modern times. Double-column numbered entries with extensive cross refs. and a full index (pp. 626-661). A virtual roadmap to guide the searcher among the byways of science and technology. Although not classified by subject or field, some entries are grouped; for instance, Marconi, Popov, Fessenden, Braun and Pupin are together under 387 to 387d, and these contain 21 refs. to related entries.

356 Debus, Allen G. (ed.), World Who's Who in Science. A Biographical Dictionary of Notable Scientists from Antiquity to the Present. Chicago: A.N. Marquis, 1968. 1855 pp. Full details of education, vocation, discoveries, inventions, awards. Inventors, engineers, and other contributors to technology are included.

357 Williams, Trevor I. (ed.), A Biographical Dictionary of Scientists. London: A. and C. Black; New York: Wiley-Interscience, 1969. 592 pp. Fifty contributors cover the broad field of science, mathematics, medicine and technology. Living persons excluded. Each entry gives highlights of the subject's important work, with cross refs. Most entries have several (1-5) refs. to sources, with mention of ports.

358 Gillispie, Charles C. (chief ed.), Dictionary of Scientific Biography. New York: Charles Scribner's Sons, 1970. A new multivolume work containing short article summaries of the professional careers and scientific accomplishments of men of all periods. Includes inventors, engineers and physicists as well as prominent workers in all fields of science, except those now living. The articles, mostly one or two columns, are signed by the contributors. There is a brief bibliography of original works and biographical sources with each entry. Three vols. (A-D) have been published. Sponsored by the American Council of Learned Societies.

B. Collected Biographies

359 Appleyard, Rollo, Pioneers of Electrical Communications. London: Macmillan, 1930. 347 pp., 138 facsimiles, ports., diags. A study of 10 men: Maxwell, Ampere, Volta, Wheatstone, Hertz, Oersted, Ohm, Heaviside, Chappe, Ronalds. Chap. 1, pp. 2-30, Maxwell; Chap. 5, pp. 108-140, Hertz, particularly relate to electronics. No formal references, but pertinent names and dates are given in the text. Some of the many extracts have English translations. This scholarly work also appeared in Elect. Comm., Vols. 5-8, 1926-1929.

360 Dibner, Bern, Ten Founding Fathers of the Electrical Science. Norwalk, Conn.: Burndy Library, 1954. 46 pp., 21 ports., illus. Also, Elect. Eng., Vols. 73, 74, April 1954-Jan. 1955. Gilbert, Guericke, Franklin, Volta, Ampere, Ohm, Gauss, Faraday, Henry, Maxwell.

361 O'Reilly, Michael F. (Brother Potamian), and James J. Walsh, Makers of Electricity. New York: Fordham, 1909. 408 pp., illus.

362 Lenard, Philipp (H. S. Hatfield, trans.), Great Men of Science. New York: Macmillan, 1933. 389 pp., 24 plates. 65 brief biographical surveys that include Maxwell, Hittorf, Crookes and Hertz. Various footnotes.

363 MacDonald, David K. C., Faraday, Maxwell and Kelvin. Garden City, N. Y.: Doubleday, 1964. 143 pp., 44 plates, 28 chap. notes.

364 Crowther, James G., British Scientists of the Nine-

teenth Century. London: Kegan Paul, Trench, Trubner, 1935. 332 pp., illus. Chap. bibliographies, also a select bibliography, p. 327. Contents include Davy, Faraday, Joule, William Thomson, Maxwell. Although of some electrical interest, the treatment is in a social and political context. See also 373.

365 Crowther, James G., Statesmen of Science. London: Cresset Press, 1965. 391 pp., 9 plates, 123 chap. refs. Brief biographical survey of nine men of science who, as statesmen, affected the progress of science (and technology) in Britain from the mid-nineteenth century.

366 Schuster, Arthur, and Arthur E. Shipley, Britain's Heritage of Science. London: Constable, 1917. 334 pp., 15 plates (ports.). Early physical science and scientists to 1800, pp. 1-105. Nineteenth century, pp. 106-186. Industrial applications, pp. 187-202. Scientific institutions, pp. 203-215. 29 footnotes in these pages, dates in the text, but no other refs. Pages 216-319 treat the biological sciences.

367 Turner, Dorothy M., Makers of Science. Electricity and Magnetism. Oxford: University Press, 1927. 184 pp., 73 illus. A popular introduction that ranges from early ideas about electricity to more recent work on electricity in gases and the electrical constitution of matter. Biographical.

368 Perucca, Eligio, and Vittorio Gori, "Pioneers in Electrical Communications." J. F. I., 261:61-79, Jan. 1956. 14 footnotes with refs. from 1753. Numerous names and dates in the text. On the electric telegraph, telegraph lines, telephone, wireless, radio commounications. A brief survey with highlights of progress up to the 1920s.

369 Hawks, Ellison, Pioneers of Wireless. London: Methuen, 1927. 304 pp., 24 plates, 45 diags. List of notable dates, pp. xvii, xviii. List of notable pioneers, p. xix. 200 footnotes, mainly full refs. From Gilbert to Faraday, wire telegraphs, water and earth conduction experiments, induction methods, developments from Maxwell to Marconi, commercial wireless telegraphy and telephony. A thorough exposition with numerous quotations and extracts, mostly supported with fairly complete refs. Originally published in parts in Wireless World and Radio Review, Vols. 18, 19, 1926.

370 "Reminiscences of Old-Timers." Radio-Craft, 9:556-561, 620, 622, 624, 626, 628, March 1938. Observations by 18 well-known men. Portrait and brief biographical matter with each entry.

371 Dunlap, Orrin E., Radio's 100 Men of Science. Biographical Narratives of Pathfinders in Electronics and Television. New York: Harper and Brothers, 1944. 294 pp., 16 plates containing 96 ports. 85 footnotes. Brief personal sketches with numerous quotations and extracts. A popular and general treatment with much useful information about the subjects and their contributions, but with very little documentation. The footnotes supplement the text; only a few contain specific and complete refs.

372 Heathcote, Neils H. de V., Nobel Prize Winners in Physics, 1901-1950. New York: Henry Schuman, 1954. 473 pp., 46 diags. Biographical sketch and description of the prize-winning work of 54 physicists. Extracts from the Nobel lectures are given with brief comments on the consequences in theory and practice. No refs. to original sources. See also 213, 214, 506.

373 Crowther, James G., British Scientists of the Twentieth Century. London: Routledge and Kegan Paul, 1952. 320 pp., 8 plates. Chapter bibliographies, also a select bibliography, p. 311. Subjects of present interest include Joseph J. Thomson and Ernest Rutherford. Same context as in the companion volume for the 19th century (364).

C. Individual Biographies

This section contains most of the book-length biographies of notable men in the fields of interest. However, the life and work of certain pioneers, such as Bell, Edison, Faraday, Henry, Maxwell, and Marconi, have been so thoroughly documented that only a few of the most representative works are included. About one-third of the entries are articles or brief notices; some of these represent rather scanty records, others supplement the more extensive biographies. A fair number of rather minor items that merely repeat material contained in larger works or are too general have been excluded. Entries are in alphabetical order.

ALEXANDERSON, Ernst Fredrik Werner, b. 1878.

374 Muir, R. C., "The 1944 Edison Medalist." Elect. Eng., 64:138, 139, April 1945. Survey of E. F. W. Alexanderson's contributions to radio technology. Also "The Edison Tradition--A Challenge to Industry," pp. 139-140 by the recipient.

ARMSTRONG, Edwin Howard, 1890-1954.

375 Hazeltine, Alan, "E. H. Armstrong--Edison Medalist." Elect. Eng., 62:147-149, April 1943. Survey of Armstrong's contributions to radio technology. Also "Vagaries and Elusiveness of Invention," pp. 149-151, by the recipient. Port.

376 Lessing, Lawrence P., "Armstrong of Radio." Fortune, 37:88-91, 198, 200-202, 204, 206, 208-210, Feb. 1948. 8 photos. Survey of Armstrong and his contributions to radio engineering, with special reference to his competitive struggles and the FM story.

377 Lessing, Lawrence P., "The Late Edwin H. Armstrong." Sci. Am., 190:64-69, April 1954. 2 photos, 9 diags. A brief review of the struggles and achievements of Armstrong as an independent inventor.

378 Ragazzini, John R., "Creativity in Radio. Contributions of Major Edwin H. Armstrong." J. Eng. Ed., 45:112-119, Oct. 1954. A personal study set against the historical background by a close colleague.

379 Lessing, Lawrence P., Man of High Fidelity: Edwin Howard Armstrong. New York: Lippincott, 1956. 320 pp., 8 plates, 8 pp. diags. Short bibliography lists 26 papers by the inventor from 1913 to 1953. A solid and colorful story; popular, nontechnical, with little documentation. Reprinted, Bantam Books, 1969. No photos, but an added chapter.

380 Lessing, Lawrence P., "The Man Behind Stereo FM." HiFi/Stereo, Dec. 1961, pp. 49-52. Port., 3 photos. Brief survey of E. H. Armstrong's life and work, particularly in developing FM radio. See also 1314.

BAIRD, John Logie, 1888-1946.

381 Tiltman, Ronald F., Baird of Television. The Life Story of John Logie Baird. London: Seeley, Service, (1933). 220 pp., 8 plates and other illus. A popular account by a reporter and friend of the inventor. Noncritical and vague except for a few dates in the text.

382 Smith-Rose, Reginald L., "John Logie Baird." Nature, 158:88, 89, July 20, 1946.

383 Moseley, Sydney, John Baird. The Romance and Tragedy of the Pioneer of Television. London: Odhams, 1952. 256 pp., 9 plates. A nontechnical account giving the "inside story" by Baird's staunchest supporter. Moseley was a business writer, promoter and television publicist.

384 Percy, J.D., John L. Baird. The Founder of British Television. London: Television Society, 1952. 16 pp., 13 illus. Chronologically arranged from 1926 to 1936.

BELL, Alexander Graham, 1847-1922.

385 Mackenzie, Catherine D., Alexander Graham Bell. The Man Who Contracted Space. New York: Houghton, Mifflin; Grosset and Dunlap, 1928. 382 pp., 7 plates. Numerous extracts, but no documentation.

386 Osborne, H.S., "Alexander Graham Bell, 1847-1922." Nat. Acad. Sci., Biographical Memoirs, 23:1-29, 1945. Port. Bibliography.

387 Mann, F.J., "Alexander Graham Bell--Scientist." Elect. Eng., 66:215-229, March 1947; also, Elect. Comm., 24:2-23, March 1947. 10 illus. 11 refs. from 1908. A survey of Bell's life on the centenary of his birth.

BERLINER, Emile, 1851-1929.

388 Wile, Frederick W., Emile Berliner, Maker of the Microphone. Indianapolis: Bobbs-Merrill, 1926. 353 pp., 23 plates. Refs. in the text. Apps., pp. 309-336, 6 miscellaneous items.

BRAUN, Karl Ferdinand, 1850-1918.

389 Lewis, G., and F. J. Mann., "Ferdinand Braun--Inventor of the Cathode-Ray Tube." Elect. Comm., 25:319-327, Dec. 1948. 7 footnotes. The only account available in English. An English translation of Ferdinand Braun, by F. Kurylo (Heinz Moos, Münich, 1965), is planned. See also 885.

COOLIDGE, William David, b. 1873.

390 Miller, John A., Yankee Scientist: William David Coolidge. Schenectady: Mohawk Development Service, 1963. 216 pp., 10 full-page illus. There are about 80 chap. refs., including some unpublished material. List of published papers by Dr. Coolidge, pp. 194-199, 72 entries. List of patents, pp. 200, 201, 83 entries. List of citations for awards and medals, pp. 202-204, 16 entries. Honorary degrees and membership list, pp. 205-207.

CROOKES, William, 1832-1919.

391 d'Albe, Edmund E. Fournier, The Life of Sir William Crookes. London: Fisher Unwin, 1923. 412 pp., ports. Foreword by Sir Oliver Lodge. Biography of a pioneer investigator of electric conduction in gases. Pages 278-290, on the "fourth state of matter," furnish background details of his famous experiments on conduction phenomena in vacuum tubes. See 719, 720.

de FOREST, Lee, 1873-1961.

392 Arvin, W. B., "The Life and Work of Lee de Forest." Radio News, 6:464, 658, 912, 1154, 1402, 1636, 1870, 2066, 2212; 7:26, 152, 287, 414, 602. Oct.-Dec. 1924, Jan.-Nov. 1925. Illus.

393 "Dr. Lee de Forest Writes the Reminiscences of a Radio Pioneer." Radio News, 11:231-233, 281, 282, 284. Sept. 1929. 8 illus.

394 Carneal, Georgette, A Conqueror of Space. New York: Liveright, 1930. 296 pp., 6 plates. An au-

thorized biography of the life and work of Lee de Forest. Popular and nontechnical. No index.

395 Lubell, Samuel, "Magnificent Failure." Saturday Evening Post, 214:9-11, 75, 76, 78, 80; 20, 21, 35, 36, 38, 43; 27, 38, 40-42, 46, 48, 49; Jan. 17, 24, 31, 1942. Illus. A highly readable, colorful and revealing story of de Forest's tumultuous life and activities in business and electrical inventing.

396 de Forest, Lee, Father of Radio. Chicago: Wilcox and Follett, 1950. 502 pp., 16 plates. Appendix contains Poems, pp. 469-476; Evolution of the Audion, pp. 477-484; List of patents, 1902-1949, pp. 485-490, 208 entries. A record of fifty active years, compiled from memory and notebooks, by an early radio pioneer best known for his invention of the triode. A breezy, poetic and often flamboyant revelation of the man--his personal life, inner sentiments, philosophy, triumphs and disasters--liberally sprinkled with extracts from newspapers and personal and business correspondence. Little documentation.

397 Gernsback, Hugo, "Lee de Forest, Father of Radio 1873-1961." Radio-Electronics, 22:33, Sept. 1961.

EDISON, Thomas Alva, 1847-1931.

398 Dyer, Frank L., and Thomas C. Martin, Edison, His Life and Inventions. New York: Harper and Brothers, 1910. Vol. 1, 472 pp., 17 plates. Vol. 2, pp. 473-989, 8 plates, index pp. 973-989. This volume contains a long and useful App., pp. 785-971. Some details of 19 inventions are given in pp. 785-942, with 70 illus. List of U. S. patents, pp. 943-970, from 1868 to 1909, arranged by year with title, number and date. List of foreign patents, p. 971, gives breakdown of 1239 patents by countries. This authorized biography lacks references except for items mentioned in the text. 2nd ed., 1929.

399 Kennelly, Arthur E., "Thomas Alva Edison, 1847-1931." Nat. Acad. Sci., Biographical Memoirs, 15: 287-304, 1934. Port. Bibliography, pp. 303, 304.

400 Josephson, Matthew, Edison. New York: McGraw-Hill, 1959. 511 pp., 16 plates. Chap. refs., pp. 487-499. A thorough, serious and valuable study. There are

692 refs. to sources and original material, including that in the Edison Library, West Orange, New Jersey.

FARADAY, Michael, 1791-1867.

401 Tyndall, John, Faraday as a Discoverer. London: Longmans, Green, 1868; reprinted, New York: Crowell, 1961. 213 pp., illus.

402 Jones, Henry B., The Life and Letters of Faraday. London: Longmans, Green, 1870. 2nd ed. rev., 2 vols., 385,491 pp. Port., 8 engrs., 8 diags. The standard work. See also 643.

403 Williams, L. Pearce, Michael Faraday. London: Chapman and Hall; New York: Basic Books, 1965. 531 pp., 37 plates, other diags. in text. List of picture sources. Refs. at the end of each chap. total 930 items. Based on original materials, this deep study includes intimate details along with a balanced view of the major events, discoveries and speculations of Faraday and his contemporaries.

FARNSWORTH, Philo Taylor, 1906-1971.

404 Everson, George, The Story of Television. The Life of Philo T. Farnsworth. New York: Norton, 1949. 266 pp., 4 plates. No index. A nontechnical story ("dramatic and absorbing") for the layman by a close associate, promoter, backer and strong supporter of the inventor. Although disappointingly vague on technical facts, dates and events, and loose in historical sequence, this is the only book to date that deals with the early work of a foremost television pioneer.

FESSENDEN, Reginald Aubrey, 1866-1932.

405 Fessenden, Helen M., Fessenden--Builder of Tomorrows. New York: Coward-McCann, 1940. 362 pp. Bibliography, pp. 353-362, has over 150 classified entries (most incomplete), dealing mainly with wire and wireless telegraphy and telephony, electrical engineering, physics and mathematics. Although Fessenden took out over 500 patents, these are not listed. A personal record by Mrs.

Fessenden that reveals domestic and business affairs. Numerous extracts of correspondence and diary entries. An account of the struggles, aspirations, achievements and failures of a pioneer inventor best known for his high-frequency alternator, heterodyne theory, early broadcasting efforts and transatlantic wireless telegraphy system. No index.

FLEMING, John Ambrose, 1849-1945.

406 Fleming, John A., Memories of a Scientific Life. London and Edinburgh: Marshall, Morgan & Scott, (1934). 244 pp., port. and 2 plates of early electron tubes. Foreword by Sir Oliver Lodge. On the early days of the telephone and electric lighting, the author's work as a professor and consultant, his scientific researches, popular lectures and scientific books, along with personal events. The thermionic valve and related matters are treated in pp. 134-147, and early work with Marconi in pp. 114-125. Some names and dates are given in the text, otherwise there are no refs. No index.

407 Eccles, William H., "Sir Ambrose Fleming, F.R.S." Nature, 155:662, 663, June 2, 1945.

408 MacGregor-Morris, John T., "Birth Centenary of Ambrose Fleming." Electronic Engineering, 21:442-447; 22:22-26, Dec. 1949, Jan. 1950.

409 MacGregor-Morris, John T., The Inventor of the Valve. A Biography of Sir Ambrose Fleming. London: Television Society, 1954. 141 pp., 22 illus. App. 1, Historical survey of electrical science (and Fleming's life) from 1790 to 1954, pp. 114-123. App. 2, Some personal recollections of Sir Ambrose Fleming, by Arthur Blok, pp. 124-134. App. 3, Chronological list of Fleming's technical books in the library of the Institution of Electrical Engineers, from 1888 to 1943, p. 135, 33 items. A concise biography by an early associate of Dr. Fleming.

HEAVISIDE, Oliver, 1850-1925.

410 Lee, G., Oliver Heaviside. London: Longmans, Green, 1947. 32 pp., 6 illus. (Science in Britain Series.)

411 The Heaviside Centenary Volume. London: Institution of Electrical Engineers, 1950. 98 pp., 6 plates, other illus. 15 papers, with footnotes, refs., list of papers and a bibliography.

HELMHOLTZ, Hermann Ludwig Ferdinand von, 1821-1894.

412 Königsberger, Leo (Frances A. Welby, trans.), Hermann von Helmholtz. Oxford: Clarendon Press, 1906. 440 pp. Ports. Preface by Lord Kelvin. Reprinted, New York: Dover, 1965.

413 Crombie, A. C., "Helmholtz." Sci. Am., 198:94-96, 98, 100, 102, March 1958. 2 ports., 2 photos, 1 diag.

HENRY, Joseph, 1797-1878.

414 [U. S.]. A Memorial of Joseph Henry. Washington, G. P. O., 1880. 528 pp., port. Obsequies, pp. 7-34, Memorial exercises, pp. 37-122. Memorial proceedings, pp. 125-508, 11 addresses, discourses and memoirs. "The Scientific Work of Joseph Henry," by William B. Taylor, pp. 205-364; Notes, pp. 375-425. List of Henry's scientific papers, pp. 365-374.

415 Gherardi, Bancroft, and Robert W. King, "Joseph Henry, the American Pioneer in Electrical Communication." B. S. T. J., 5:1-10, Jan. 1926. Port., 2 photos. 3 footnotes.

416 Coulson, Thomas, Joseph Henry, His Life and Work. Princeton, N. J.: Princeton University Press, 1950. 352 pp., port., 7 plates. Bibliography, pp. 344-346, 45 entries. 266 footnote refs.

HERTZ, Heinrich Rudolph, 1857-1894.

417 Hertz, Heinrich R. (Daniel E. Jones and George A. Schott, trans.), Miscellaneous Papers. London, New York: Macmillan, 1896. 340 pp., illus. Introduction by Philipp Lenard.

418 Blanchard, Julian, "Hertz, the Discoverer of Electric Waves." Proc. IRE., 26:505-515, May 1938. Also

B.S.T.J., 17:327-337, July 1938. No refs.

419 Morrison, Philip, and Emily, "Heinrich Hertz." Sci. Am., 197:98-100, 102, 104, 106, Dec. 1957. Port., 3 diags.

HUGHES, David Edward, 1831-1900.

420 Evershed, Sydney, "The Life and Work of David Hughes." J. IEE., 69:1245-1250, Oct. 1931. Also, Nature, 127:822-824, May 30, 1931. From an address to the Institution of Electrical Engineers.

HULL, Albert Wallace, 1880-1966.

421 Suits, C.G., and J.M. Lafferty, "Albert Wallace Hull, 1880-1966." Nat. Acad. Sci., Biographical Memoirs, 41:215-233, 1970. Port. Bibliography, pp. 228-233.

IVES, Herbert Eugene, 1882-1953.

422 Buckley, Oliver E., and Karl K. Darrow, "Herbert Eugene Ives." Nat. Acad. Sci., Biographical Memoirs, 29:145-189, 1956. Port. Bibliography, pp. 174-186, Patents, pp. 187-189.

JENKINS, Charles Francis, 1867-1934.

423 Jenkins, Charles F., The Boyhood of an Inventor. Washington: The Author, 1931. 273 pp., 252 illus. 8 broadcast talks by the author, pp. 205-272. A revealing record by a man who regarded himself as an inventor above all else. Founder of the Society of Motion Picture Engineers and pioneer in phototelegraphy and radio pictures, he deals with these developments in pages 122 to 167. No index.

KENNELLY, Arthur Edwin, 1861-1939.

424 Bush, Vannevar, "Arthur Edwin Kennelly, 1861-1939." Nat. Acad. Sci., Biographical Memoirs. 22:83-119, 1943. Port. 17 footnotes. Bibliography, pp. 95-119.

KORN, Arthur, 1870-1945.

425 Korn, Terry, and Elizabeth P. Korn, Trailblazer to Television. New York: Charles Scribner's Sons, 1950. 144 pp., 6 full-page drawings by Elizabeth Korn. A popular story of the life and work of Arthur Korn by his wife and daughter-in-law. Undocumented, no index.

LANGMUIR, Irving, 1881-1957.

426 Suits, C. Guy (ed.), The Collected Works of Irving Langmuir. New York: Pergamon Press, 1962, 12 vols. Vol. 12, The Man and The Scientist, 473 pp. Pt. 1, "The Quintessence of Irving Langmuir: A Biography," by Albert Rosenfeld, pp. 3-225. List of principal sources, pp. 226-229, 133 items. App. 1, pp. 459, 460, list of scholastic dates, degrees, medals. App. 2, pp. 461-473, complete listing of Langmuir's papers contained in Vols. 1-12. No index. Also published separately, 1966, 369 pp.

LODGE, Oliver Joseph, 1851-1940.

427 Lodge, Oliver J., Past Years, An Autobiography. London: Hodder and Stoughton, 1931; New York: Charles Scribner's Sons, 1932. 364 pp., 13 plates. Chap. 17, pp. 225-236, deals with electric waves and the beginnings of wireless. A short list of salient dates about ether waves, pp. 234-236, has 31 entries from 1864 to 1922. Numerous refs. in the text. The author was a prominent English physicist and leading radio pioneer.

428 Gregory, Richard A., and A. Ferguson, "Sir Oliver Joseph Lodge, 1851-1940." Roy. Soc., Obituary Notices of Fellows, 3:551-574, 1939-1941.

LOOMIS, Mahlon, 1826-1886.

429 Appleby, Thomas, Mahlon Loomis, Inventor of Radio. Washington: Loomis Publications, 1967. 145 pp., plus 19 p. index, 37 illus. Copious extracts from contemporary sources with some refs. See also 1222.

LORENTZ, Hendrik Antoon, 1853-1928.

430 "Hendrik Antoon Lorentz." Nature, 121:287-291, Feb. 25, 1928. Obituary notices by 10 contributors.

MARCONI, Guglielmo, 1874-1937.

431 Jacot, Bernard L., and D. M. B. Collier, Marconi, Master of Space. London: Hutchinson, 1935. 287 pp., 16 plates. List of honors, pp. 279-281. An authorized biography, popular and nontechnical.

432 Dunlap, Orrin E., Marconi, The Man and His Wireless. New York: Macmillan, 1937. 360 pp., 16 plates. Two-page list of honors, degrees, awards. Numerous quotations and extracts, a few footnotes and occasional text refs. A popular and somewhat romantic survey in which Marconi co-operated. Based on much original material, but lacks specific refs.

433 Coe, D., Marconi, Pioneer of Radio. New York: Julian Messner, 1943. 272 pp., 17 diags. List of Marconi's awards. Bibliography contains 39 items; books, periodical and newspaper articles.

434 Marconi, Degna, My Father, Marconi. New York: McGraw-Hill, 1962. 321 pp., 16 plates. A family record, based on personal notes, journals and scientific records, by the famous inventor's eldest daughter. Interesting and useful for background material. Includes a variety of extracts, but has few refs.

MAXWELL, James Clerk, 1831-1879.

435 Tait, Peter G., "Clerk-Maxwell's Scientific Work." Nature, 21:317-321, 1879-1880.

436 Campbell, Lewis, and William Garnett, The Life of James Clerk Maxwell. London: Macmillan, 1882. With selections from his correspondence and occasional writings. New revised and abridged ed., 1884. 421 pp., ports., illus., and diags.

437 Niven, William D. (ed.), The Scientific Papers of

James Clerk Maxwell. Cambridge: University Press, 1890; New York: Dover, 1965. 2 vols. in 1, pp. 607, 806. Index in both vols. Plates, diags., tables. Ports.; Maxwell, Faraday, Helmholtz.

438 Jones, E. T., "The Life and Work of James Clerk Maxwell, 1831-1879." Proc. Roy. Phil. Soc. Glasgow, 60:54-77, 1931-1932.

439 Smith-Rose, Reginald L., James Clerk Maxwell, F. R. S. 1831-1879. London: Longmans, Green, 1948.

440 Dibner, Bern, "James Clerk Maxwell," IEEE Spectrum, 1:50-56, Dec. 1964. 3 illus., 9 refs. from 1856. A tribute on the centennial of Maxwell's historic paper "A Dynamical Theory of the Electromagnetic Field."

MILLIKAN, Robert Andrews, 1868-1953.

441 Millikan, Robert A., Autobiography. New York: Prentice-Hall, 1950. 311 pp., 8 plates. Two parts are of particular electronics interest; Chap. 7, pp. 68-86, on his "Oil-Drop Venture," and Chap. 9, pp. 100-107, on the photon. 58 footnote refs.

442 DuBridge, Lee A., and Paul S. Epstein, "Robert Andrews Millikan, 1868-1953." Nat. Acad. Sci., Biographical Memoirs, 33:241-282, 1959. Port. Bibliography, pp. 270-282.

MORSE, Samuel Finley Breese, 1791-1872.

443 Prime, Samuel I., The Life of Samuel F. B. Morse, Inventor of the Electro-Magnetic Recording Telegraph. New York: D. Appleton, 1875. 776 pp., illus., ports., diags.

444 Morse, Edward L. (ed.), Samuel F. B. Morse. His Letters and Journals. Boston; New York: Houghton, Mifflin, 1914. 2 vols., 440, 548 pp; 13, 15 plates. Index, Vol. 2, pp. 523-548.

PIERCE, George Washington, 1872-1956.

445 Saunders, Frederick A., and Frederick V. Hunt,

"George Washington Pierce, 1872-1956." Nat. Acad. Sci., Biographical Memoirs, 33:351-380, 1959. Port. Bibliography, pp. 371-374. U.S. Patents, pp. 374-380.

POPOV, Alexander Stepanovich, 1859-1905.

446 Radovsky, M. (G. Yankovsky, trans.), Alexander Popov: Inventor of Radio. Moscow: Foreign Languages Publishing House, 1957. 130 pp., several plates of family groups, ports., and others showing Popov's early instruments. This story of Russia's radio pioneer provides useful and interesting material not otherwise available.

PUPIN, Michael Idvorsky, 1858-1935.

447 Pupin, Michael I., From Immigrant to Inventor. New York; London: Charles Scribner's Sons, 1922; popular ed., 1925. 396 pp., 7 plates. A popular autobiography, nontechnical, no refs.

448 David, B., "Michael Idvorsky Pupin, 1858-1935." Nat. Acad. Sci., Biographical Memoirs, 19:307-323, 1938. Port. Bibliography, pp. 319-323.

RÖNTGEN, Wilhelm Konrad, 1845-1923.

449 Glasser, Otto, Wilhelm Conrad Röntgen and the Early History of the Roentgen Rays. Springfield, Ill.: Charles C. Thomas, 1934. 494 pp., 96 illus. Bibliography, pp. 421-479, 1044 entries plus a list of articles concerning W.C. Röntgen, 34 items. Basic source book on Röntgen's life and work, particularly on early X-ray history before 1900. Port. See 692.

450 Glasser, Otto, Dr. W.C. Röntgen. Springfield, Ill.: Charles C. Thomas, 1945. 2nd ed., 1958. 169 pp., 20 plates and facsimiles. Scientific papers of W.C. Röntgen, pp. 145-149, 58 items from 1870 to 1921. Chronology, pp. 150-157, from 1845 to 1932. Bibliography, pp. 158-160, 21 items to 1932 plus 18 papers and books published since 1945. Early history, pp. 3-32. Events from November 1895 to March 1897 occupy most of the book; pp. 33-115. Events from 1897 to 1923, pp. 116-142.

SWINTON, Alan Archibald Campbell, 1863-1930.

451 Swinton, Alan A. Campbell, Autobiographical and Other Writings. London: Longmans, Green, 1930. 181 pp., 18 plates. A fragile and modest story of a man who accomplished far more than the text indicates. The first proposal for an all-electronic television system was put forward by Swinton in 1908 (see 1728-1730). See also 1743.

TESLA, Nikola, 1857-1943.

452 Martin, Thomas C., "Nikola Tesla." Century Magazine, 25(NS):582-585, Feb. 1894. A biographical sketch with port.

453 Martin, Thomas C., The Inventions, Researches, and Writings of Nikola Tesla. With special reference to his work in polyphase currents and high potential lighting. New York: The Electrical Engineer, 1894. 496 pp., port., 313 photos and diags. Pt. 2, pp. 119-406, The Tesla effect with high frequency and high potential currents. Includes high-frequency alternators, Tesla coil, electric discharge in vacuum tubes, with details of demonstrations, apparatus and effects.

454 O'Neill, John J., Prodigal Genius: The Life of Nikola Tesla. New York: Ives, Washburn, 1944. 326 pp., port. Includes a partial list of patents.

THOMSON, Elihu, 1853-1937.

455 Compton, Karl T., "Elihu Thomson, 1853-1937." Nat. Acad. Sci., Biographical Memoirs, 21:143-179, 1940. Port. Bibliography, pp. 163-179.

456 Woodbury, David O., Beloved Scientist: Elihu Thomson. A Guiding Spirit of the Electrical Age. New York: McGraw-Hill, 1944. 358 pp., 16 plates. Bibliography, p. 349, lists 14 items. Page 251 lists 17 medals and decorations from 1888 to 1935. A popular and readable account with interesting background material, but little technical matter.

THOMSON, Joseph John, 1856-1940.

457 Thomson, Joseph J., Recollections and Reflections. London; New York: Macmillan, 1937. 451 pp., 10 plates, 6 diags. in the text. App., pp. 435-438, list of Thomson's Cavendish students who became Fellows of the Royal Society. Chaps. 11, 12, pp. 325-371, 372-433, on discharges in gases, discovery of the electron, and contemporary physics are particularly relevant. 16 footnotes, no refs.

458 Rayleigh, Lord, Joseph John Thomson, 1856-1940. London: Royal Society, 1941. 23 pp., port., list of writings. Also, Obituary Notices of Fellows of the Royal Society, 3:587-609, 1939-1941. (See next item.)

459 Rayleigh, Lord, The Life of Sir J. J. Thomson, O. M. Sometime Master of Trinity College, Cambridge. Cambridge: University Press, 1942; reprinted, London: Dawsons of Pall Mall, 1969. 299 pp., 8 plates. App. 1, pp. 288-291, List of distinctions. App. 2, p. 292, Supplement to bibliography, 5 items overlooked in the list in Roy. Soc. Obituary Notices (preceding item). Includes many letters, extracts and quotations.

460 Price, D. J., "Sir J. J. Thomson, O. M., F. R. S. A Centenary Biography." Discovery, 17:494-502, Dec. 1956. 6 photos, 6 refs.

461 Thomson, George P., J. J. Thomson and the Cavendish Laboratory in His Day. London: Nelson, 1964; New York: Doubleday, 1965. 186 pp., 20 plates, 25 illus. Also, Discoverer of the Electron. Anchor Books, 1966. 215 pp., 12 plates, 25 illus., several footnotes. Brief App. on vacuum pumps, pp. 201-208. A concise story of a leading English physicist who discovered the electron during his pioneer work on the conduction of electricity in gases.

WATSON-WATT, Robert Alexander, b. 1892.

462 Watson-Watt, Robert A., Three Steps to Victory. London: Odhams Press, 1957. 480 pp., illus., photos. This autobiography is centered on the development of radar in Britain and the author's contributions. Several

plates show early apparatus. Details of government policies, personalities, events and technical matters, before and during World War II, interspersed with philosophical observations. Also, New York: Dial Press, The Pulse of Radar: The Autobiography of Sir Robert Watson-Watt, 1959. 438 pp. A shortened version edited for American readers, without photos, no index. App., 427-434, contains the government memorandum on the detection and location of aircraft by radio methods.

WESTON, Edward, 1850-1936.

463 Woodbury, David O., A Measure for Greatness. A Short Biography of Edward Weston. New York: McGraw-Hill, 1949. 230 pp., 19 illus., ports. A popular text, no refs.

ZWORYKIN, Vladimir Kosma, b. 1889.

464 Zworykin, Vladimir K., "The Early Days: Some Recollections." Television Quarterly, 1:69-73, Nov. 1962. On his association with Boris Rosing, early work on camera tubes at Westinghouse Electric and Manufacturing Company, and later events up to 1931.

5. GROUP HISTORIES

The history of an institution or a society mirrors activities and events that constitute progress of the respective science or art. Such a record is also a collective biography. These histories are sources of information concerning publications, professional and engineering standards, the growth of technical education and special group activities such as conventions and exhibitions. Histories of commercial organizations record leading events in research and industrial development, progress in production and services, and also tell the stories of people who contributed to these advances. This chapter includes a few general references as well as separate listings of professional and commercial organizations. See also Sect. 3G.

A. Institutional References

465 Becker, Bernard H., Scientific London. London: Henry S. King, 1874; reprinted, Frank Cass, 1968. 340 pp. Deals with the Royal Society, Royal Institution, Society of Arts, Department of Science and Art, London Institution, Society of Telegraph Engineers, among other societies. Some early electrical history. No refs., no index.

466 [British Council]. Scientific and Learned Societies of Great Britain. London: George Allen and Unwin, 1956. 58th ed., 211 pp. This "handbook compiled from official sources" is a continuation of the Yearbook of Scientific and Learned Societies which ceased publication in 1939. General science, pp. 35-44. Mathematical and physical, pp. 47-50. Engineering & architecture, p. 99-115. Select bibliography, pp. 26-29, 64 entries; periodicals, reports, books and governmental publications. Entries provide names, dates and addresses, with some details of objects, membership, fees, meetings, publications, and in some cases library collections. 60th ed., 1962.

467 Thompson, J.D. (ed.), Handbook of Learned Societies and Institutions--America. Washington: Carnegie In-

stitution, 1908. 592 pp. Covers all America, includes trade associations, institutes and clubs, technical as well as scientific, with full bibliographies for most entries.

468 Bates, Ralph S., Scientific Societies in the United States. New York: John Wiley; London: Chapman and Hall, 1945. 246 pp. A publication of the Technology Press, M. I. T. Scientific societies in eighteenth century America, national growth, 1800-1865; the triumph of specialization, 1866-1918; American scientific societies and world science, 1919-1944; the increase and diffusion of knowledge. Bibliography, pp. 193-220; lists of societies, books, periodicals, international references, histories, other references to biographical guides, American scientific agencies and history of science in America. 2nd ed., Columbia University Press, 1958. 297 pp. With an additional chapter on the atomic age, 1945-1955. Extended bibliography, pp. 225-268. 3rd ed., Cambridge, Mass.: The M. I. T. Press, 1965. 326 pp. With an additional chapter on scientific societies in the space age, 1955-1965, pp. 224-236. A new feature is a chronology of science and technology in America (from 1000? to 1965), pp. 237-244, about 280 entries. Extensive bibliography, pp. 245-293.

469 [National Academy of Sciences - National Research Council]. Scientific and Technical Societies of the United States and Canada. Washington: 1961, 7th ed. Pt. 1 (U. S.), 413 pp., 1597 items. Pt. 2 (Canada), 54 pp., 239 items. Each entry includes history, purpose, membership, meetings and, in some cases, professional activities and publications. A comprehensive listing of current information. Separate indexes.

470 Argles, Michael, South Kensington to Robbins; An Account of English Technical and Scientific Education since 1851. London: Longmans, Green, 1964. 178 pp. Landmarks in the development of technical and scientific education in England, pp. 148-150, 76 entries from 1563. Select list of books, etc., pp. 151-161. A survey from the year of the Great Exhibition to the Robbins Report on higher education, 1963-64.

471 Ferguson, Eugene S., "Technical Societies, Education, and Exhibitions." Bibliography of the History of Technology, Chap. 12, pp. 173-201. Technical and scientific societies, technical colleges and institutes, technical museums, exhibitions, diffusion of technology. About 170 en-

tries, some with detailed listings and multiple items. See 40.

B. Societies and Institutions

Entries in this section are arranged alphabetically.

American Institute of Electrical Engineers

472 Mailloux, C. O., "The Evolution of the Institute and of Its Members." Trans. AIEE., 33:819-838, 1914.

473 "Fortieth Anniversary Celebration of the AIEE." Trans. AIEE., 43:104-115, 1924.

474 Ryan, H. J., "A Generation of the American Institute of Electrical Engineers--1884-1924." Trans. AIEE., 43:740-744, 1924.

475 "Fiftieth Anniversary of the AIEE." Elect. Eng., 53:641-848, May 1934. See 263 and next item.

476 Scott, Charles F., "The Institute's First Half Century." Elect. Eng., 53:645-670, May 1934. 7 illus., including 2 showing some well-known signatures.

477 Funk, N. E., "The First 60 Years of the AIEE." Elect. Eng., 63:323-325, Sept. 1944.

478 Hooven, Morris D., "The Electrical Engineering Profession in the Past Century." Elect. Eng., 71:973-977, Nov. 1952. Highlights of the AIEE.

American Institute of Physics

479 Barton, H. A., "The Story of the American Institute of Physics." Physics Today, 9:56-58, 60, 62, 64, 66, Jan. 1956. Covers 25 years with reference to member societies and publications of the Institute.

American Radio Relay League

480 "ARRL's 40th Anniversary." QST, 38:9-12, May 1954.

5 photos. A brief review of highlights in the progress of amateur radio and the development of the League and its journal.

481 "An Anniversary Look at QST." QST, 49:9-24, Dec. 1965. 9 illus. Survey of 50 years.

British Association for the Advancement of Science

482 Howarth, O. J. R., The British Association for the Advancement of Science: A Retrospect 1831-1931. London: The Association, 1931. First published in 1922, this is the Centenary Edition. 330 pp., 21 plates, ports., views, facsimiles. App. 2, pp. 293-322, gives dates and places of annual meetings, notes of the Presidents, and additional biographical notes.

483 Howarth, O. J. R., "The British Association." Endeavour, 3:57-61, April 1944. 4 photos.

British Institution of Radio Engineers (Institution of Electronic and Radio Engineers)

484 A Twentieth Century Professional Institution. The Story of the Brit. I. R. E. London: The British Institution of Radio Engineers, 1960. 95+24 pp., colored frontispiece of the Armorial Bearings, one folding plate, 5 diags., 2 illus. Apps. 1-6, pp. ix-xviii, membership qualifications and institutional relationships. Chronological table, pp. xix-xx, 59 entries from 1871 to 1951. Education and training, services, conventions, membership, awards, administration, publications and related matters. Became Institution of Electronic and Radio Engineers in 1961.

Cavendish Laboratory (University of Cambridge)

485 Wood, A., "History of the Cavendish Laboratory, Cambridge." Endeavour, 4:131-135, 1945.

486 Larsen, Egon, The Cavendish Laboratory. Nursery of Genius. London: Edmund Ward, 1962. 95 pp. 12 plates, 3 diags. A popular story with slight technical treatment. No refs., no index.

Department of Scientific and Industrial Research (G. B.)

487 Crowther, James G., and Richard Whiddington, Science at War. London: H. M. S. O., 1947. 185 pp., illus., plates. On developments and applications of science and technology during World War II, and the role of the Department of Scientific and Industrial Research.

488 Melville, Harry, The Department of Scientific and Industrial Research. London: Allen and Unwin, 1962. 200 pp. Description of the Department since its founding in 1915, its activities, past events, associations and organization. Individual establishments mentioned include the National Physical Laboratory, pp. 101-110; and the Radio Research Station, pp. 146-150.

Franklin Institute

489 Houston, Edwin J., "The Seventy-Fifth Anniversary of the Franklin Institute from an Electrical Standpoint." J. F. I., 148:346-365, 1899.

490 Thomson, Elihu, "Centenary of the Founding of the Franklin Institute." J. F. I., 198:581-598, 1924.

Institution of Electrical Engineers

491 "The Jubilee of the Institution of Electrical Engineers." Nature, 109:284, 285, 1922.

492 Atkinson, L. B., "Institute Recollections." J. IEE., 85:553-557, Oct. 1939.

493 Appleyard, Rollo, The History of the Institution of Electrical Engineers, 1871-1931. London: The Institution, 1939. 342 pp., 86 ports., views, etc., on 36 plates. 8 Apps., pp. 301-324. Royal Charter, p. 301. List of Presidents, pp. 286-300, 57 entries, with brief biographies. History of electrical engineering, with technical and biographical details, and numerous extracts.

494 "The Wireless Section of the I. E. E. 25th Anniversary." J. IEE., 91 (Pt. 3):100-104, Sept. 1944. 2 plates. Also, Electrical Review, 134:661-664; Electrician,

132:408-410, May 12; Engineering, 157:373, 374; Electronic Engineering, 17:9-11, 1944.

495 Dunsheath, Percy, "Professional Organization." A History of Electrical Engineering, Chap. 19, pp. 319-332. Brief history of the Institution of Electrical Engineers; its organization, activities, library and publications. See 273.

Institution of Post Office Electrical Engineers

496 "50th Anniversary Issue." J. POEE., Vol. 49, Oct. 1956. There is one paper on the fifty-year history of the Institution, and 14 others that also include some history.

Institute of Radio Engineers

497 Dudley, Beverly, "Silver Anniversary--The Institute of Radio Engineers." Electronics, 10:15-21, May 1937.

498 "40th Anniversary Issue." Proc. IRE., 40:514-524, May 1952. "The Genesis of the IRE," pp. 516-520. 3 illus., one table of membership and publications. Other papers: "The Founders of the IRE," "Life Begins at Forty," "The IRE in Cohesion or Depression," "A Look at the Past Helps to Guess at the Future in Electronics."

499 Whittemore, Laurens E., "The Institute of Radio Engineers--Forty-Five Years of Service." Proc. IRE., 45:597-635, May 1957. 15 photos, 2 graphs, 11 tables. 2 Apps., publications and records, 16 categories. A full survey of the Institute's history.

500 Whittemore, Laurens E., "The Institute of Radio Engineers--Fifty Years of Service." Proc. IRE., 50:534-554, May 1962. 4 illus. App. lists 13 previous papers related to IRE history. List of IRE Officers 1912-1962. Lists and charts showing growth of membership and publications.

Massachusetts Institute of Technology, Radiation Laboratory

501 DuBridge, Lee A., "History and Activities of Radiation Laboratory of M.I.T." Rev. Sci. Inst., 17:1-5, Jan. 1946.

502 Baxter, James P., Scientists Against Time. Numerous references to the M.I.T. Radiation Laboratory during World War II. See 507.

503 Henney, Keith (ed.), Index, MIT Radiation Laboratory Series, Vol. 28, 1953. 42+160 pp. The prefatory pages contain three historical summaries: Establishment of the Radiation Laboratory, Karl T. Compton (xiii-xv); Organization of the Radiation Laboratory, Lee A. DuBridge (xvi-xxiv); Preparation of the Radiation Laboratory Series, Louis N. Ridenour (xxv-xxx). There is also a Foreword by Vannevar Bush (v-vii), and the Editor's Introduction (xxxi, xxxii). See 798.

National Academy of Sciences

504 [National Academy of Sciences]. A History of the First Half-Century of the National Academy of Sciences 1863-1913. Washington: The Academy, 1913.

National Bureau of Standards (U.S.)

505 Cochrane, Rexmond C., Measures for Progress. A History of the National Bureau of Standards. Washington: The Bureau, 1966. 703 pp., 126 illus., 2 folding charts, 1252 footnotes. 15 Apps., pp. 515-672. App. K, pp. 631-648, NBS publications representing research highlights in science and technology, 1901-1951, lists over 70 items on electricity, radio and related topics. App. N, pp. 663-667, Books by staff members, 1912-1960, has 78 entries with about a dozen on radio and electronics. Bibliographic note, pp. 673-681, has 131 cited entries that cover a wide range of subjects. Some 60 pages treat topics of present interest; early radio research and instruments, World War I radio developments, radio broadcasting, aircraft radio and instrument landing, radio beacons, radio research of the 1930s, the radiosonde, the proximity fuse, printed circuits, radar, radio propagation, computers.

Nobel Foundation

506 Nobel. The Man and His Prizes. Edited by the Nobel Foundation. Norman, Oklahoma: University of Oklahoma Press, 1951. 620 pp. An authoritative account of Alfred Nobel, the Foundation, the prizes and prize-winners, with background material of the subject fields. List of prize-winners, pp. 605-607. Name index only. 2nd ed., 1962, 690 pp. List of prize-winners, pp. 666-673, to 1961. A chapter on the physics prize is listed separately, see 214. See also 213, 372.

Office of Scientific Research and Development (U. S.)

507 Baxter, James P., Scientists Against Time. Boston: Little, Brown, 1946. 473 pp., 63 plates, 3 diags. This official story of the Office of Scientific Research and Development also describes the origin, growth and activities of the M. I. T. Radiation Laboratory and the work of associated groups and contractors. Parts of particular interest include radar and loran (136-157), radar countermeasures (158-169), fire control (212-220), proximity fuses (221-242). Names and dates in the text, occasional footnote references, no other documentation.

Radio Club of America

508 Fiftieth Anniversary Golden Yearbook. New York: Radio Club of America, 1959. 216 pp., many photos. Official history of the Radio Club, with statistics, lists of members, awards, papers, etc.

Radio Society of Great Britain

509 "History of the Wireless Society of London." Wireless World, 11:257-263, 1922.

510 "Golden Jubilee Issue." RSGB Bull., Vol. 39, July 1963. See 1331, 1333.

Royal Institution

511 Martin, Thomas, The Royal Institution. London, New York: Longmans, Green, 1948. 52 pp., plates, ports. Short bibliography. Published for the British Council.

Royal Society

512 Stimson, Dorothy, Scientists and Amateurs. A History of the Royal Society. New York: Henry Schuman, 1948. 270 pp., 17 plates. Bibliography and sources, pp. 251-262, 136 entries grouped by chapters. "The first authoritative book for the general reader."

Royal Society of Arts

513 Wood, Henry T., A History of The Royal Society of Arts. London: John Murray, 1913. 558 pp., 31 plates, 19 text illus. Apps., pp. 509-520; the Society's officials (1754-1913), the Albert Medal (1864-1913), portraits in the Society's possession.

514 Hudson, Derek, and Kenneth W. Luckhurst, The Royal Society of Arts, 1754-1954. London: John Murray, 1954. 411 pp., 25 plates, 9 text illus. Apps., pp. 372-384; the Society's officials (1754-1953), the Albert Medal, publications; over 100 entries of serials, books and reports. Electricity and magnetism, the telegraph and telephone, wireless telegraphy, television and the magnetron are mentioned.

Science Museum (London)

515 Taylor, F. Sherwood, "The Science Museum, London." Endeavour, 10:82-88, April 1951. 2 plates.

516 The Science Museum. The First Hundred Years. London: H. M. S. O., 1957. 85 pp., 38 illus., mostly full-page. Brief account of the Museum's history, pp. 1-10, with some details of 21 exhibits. Fleming's original wireless valve, pp. 74-76, with photo.

Smithsonian Institution (U. S. National Museum)

517 Goode, George B. (ed.), The Smithsonian Institution 1846-1896. The History of Its First Half Century. Washington, 1897. 856 pp. App., pp. 833-839, Principal events in the history of the Institution from 1826, compiled by William J. Rhees.

518 The Smithsonian Institution. Washington: The Institution, 1965. 128 pp. 181 illus., many full-page and many in color. Popular survey

519 Oehser, Paul H., The Smithsonian Insitution. New York: Praeger, 1970. 275 pp. 8 plates, map, organization chart. Selected Smithsonian publications, pp. 240-247, 101 entries. Bibliography, pp. 264-268, 70 entries.

Society of Motion Picture and Television Engineers

520 Jenkins, Charles F., "Society History." Trans. SMPE., 7:6-8, Nov. 1918. See also 423.

521 Matthews, Glenn E., "Historic Aspects of the SMPTE." J. SMPTE., 75:856-867, Sept. 1966. 11 ports., 6 photos, 2 illus. Extensive bibliography, pp. 862-867, 37 items in 3 categories; the work of pioneers, general historical papers to 1965, technical papers to 1940. About two dozen entries are concerned with the history of sound films and television. See also 423.

C. Commercial Organizations

Entries in this section are arranged alphabetically.

American Telephone and Telegraph Company

522 Coon, Horace, C., American Tel. and Tel. The Story of a Great Monopoly. New York: Longmans, Green, 1939. 276 pp. A general history from the invention of the telephone. Also deals with the telegraph, radio, and sound-recording. Bibliography, pp. 267-272.

523 Danielian, N. R., A. T. and T. The Story of Indus-

trial Conquest. New York: Vanguard Press, 1939. 460 pp. Primarily a business history. Three chaps. are related to technical aspects; 5, pp. 99-119, Science in business; 6, pp. 120-137, Science at work; 7, pp. 138-172, The Radio-Bell Axis. These deal with inventions, patents, litigation and agreements related to early electron tubes, circuits, wire and radio communications, broadcasting, sound motion pictures, products and services. 578 chap. refs. See also 337, 1608.

Bell Telephone System and Laboratories

524 Findley, P. B., "Our Historical Museum." Bell Lab. Rec., 1:137-146, 1925-26. Historical telephone apparatus.

525 Farnell, William C. F., The Bell System Historical Museum. New York: Bell Telephone Laboratories, 1936. 50 pp., illus. Description of historical apparatus and guide to the exhibits. Also, Bell Tel. Qtly., 15:169-187, 261-273, 1936.

526 Page, Arthur W., The Bell Telephone System. New York, London: Harper and Brothers, 1941. 248 pp., 43 charts, maps, tables and diags. On company management, policies, practices, personnel, activities, services and growth, with some historical details. There are chapters on research, Bell Laboratories, Western Electric Company, and public utility regulation. Bell System statistics, pp. 209-212.

527 Bello, Francis, "The World's Greatest Industrial Laboratory." Fortune: 58:148-157, 208, Nov. 1958. On the Bell Telephone Laboratories. 47 illus. A picture survey with supporting text on men, inventions, products, activities and historic events.

528 Doherty, W. H., "The Bell System and the People Who Built It." Bell Lab. Rec., 46:38-46, 76-83, Feb., March, 1968. 16 photos, diags. and facsimiles. A resumé of telephone history and growth of the Bell System "based on an orientation talk given to new technical employees at Bell Laboratories."

British Thomson-Houston Company

529 Price-Hughes, H. A. (comp.), B. T. H. Reminiscences. Sixty Years of Progress. Rugby: British Thomson-Houston Co., 1946. 176 pp., illus.

A. C. Cossor Limited

530 Half a Century of Progress. London: A. C. Cossor, 1947. 32 pp., over 50 illus. from 1896. Cathode-ray tubes and oscilloscopes, X-ray tubes, electron tubes, radio and television receivers, radar, military equipment.

Du Mont Laboratories

531 Pioneering the Cathode-Ray and Television Arts. Passaic, N. J.: Allen B. Du Mont Laboratories, 1941. 32 pp., about 50 illus. A report of the first decade of the Du Mont organization. Cathode-ray tubes, oscilloscopes, television transmitters and receivers. List of Du Mont patents.

Edison Swan Electric Company

532 The Pageant of the Lamp. London: The Edison Swan Electric Co., c. 1939. 72 pp., over 50 illus. "The story of the electric lamp" from early experiments of Joseph Swan, Edison's work and their joint company, to the rise of electric lighting and its growth in Britain.

Federal Telephone and Radio Corporation

533 Mann, F. J., "Federal Telephone and Radio Corporation. A Historical Review: 1909-1946." Elect. Comm., 23:376-405, Dec. 1946. 27 illus. 23 footnote refs. Bibliography, p. 405, 15 entries from 1913. Poulsen arc transmitters, early radio stations, Palo Alto laboratory, company organization and changes, work during World War II.

General Electric Company (U. S.)

534 Jehl, Francis, Menlo Park Reminiscences. Dearborn, Mich.: Edison Institute. 3 vols., 1937, 1938, 1941. Vol. 1, 430 pp.; Vol. 2, pp. 451-901; Vol. 3, pp. 923-1156, plus index to contents and illus. in each vol. Over 1000 illus. Chap. 12, pp. 80-88, on etheric force; Chap. 13, pp. 89-93, on Edison and the wireless. Numerous refs. in the text and in footnotes. A thorough record of events at Menlo Park up to the 1880s, written by a close associate of Edison.

535 Broderick, John T., Forty Years with General Electric. Albany, N. Y.: Fort Orange Press, 1929. 218 pp., illus. Primarily about company operations and notable individuals.

536 Hammond, John W., Men and Volts. The Story of General Electric. Philadelphia: J. B. Lippincott, 1941. 436 pp., 77 plates, 9 illus. in the text. App., pp. 395-424, statement of Owen D. Young, May 17, 1939.

537 Miller, John A., Men and Volts at War. The Story of General Electric in World War II. New York: McGraw-Hill, 1947. 272 pp., 123 plates. List of plants and principal war products, pp. 246-251, Production awards, pp. 252-255.

538 Hawkins, Laurence A., Adventure Into the Unknown. The First Fifty Yers of the General Electric Research Laboratory. New York: William Morrow, 1950. 150 pp., 26 illus. A series of sketches on the character of the laboratory and some of its projects, with emphasis upon personalities. No documentation, no index.

539 Birr, Kendall, Pioneering in Industrial Research. The Story of the General Electric Research Laboratory. Washington: Public Affairs Press, 1957. 204 pp. Chap. refs., pp. 185, 186, 45 entries. Notes on sources, pp. 187-199.

General Electric Company (G. B.)

540 Whyte, Adam G., Forty Years of Electrical Progress: The Story of the G. E. C. London: Ernest Benn, 1930.

166 pp., 22 plates, including 20 pictures of early G. E. C. products. The rise and progress of the General Electric Company.

General Radio Company

541 Sinclair, Donald B., The General Radio Company 1915-1965. New York: Newcomen Society, 1965. 32 pp., 4 plates (showing 8 instruments from 1918), 5 other illus. An address on the company history.

542 Thiessen, Arthur E., A History of the General Radio Company. West Concord, Mass.: General Radio Co., 1965. 116 pp., 25 illus. App., pp. 93-111, list of people and the company structure.

W. T. Henley's Telegraph Works Company

543 Slater, Ernest, One Hundred Years. The Story of W. T. Henley's Telegraph Works Company. London: Henley's Telegraph Works, 1937. 77 pp.

Marconi Company

544 Leigh-Bennett, Ernest P., A City of Sound. London: Marconiphone, 1933. 49 pp., 8 plates. Story of Marconi Company and the factory at Hayes, Middlesex.

545 Jubilee Year. Chelmsford: Marconi Co., 1947. 57 pp., port. of G. Marconi, 51 illus. Includes Marconi patent 12,039, and indenture for British Empire chain, July 28, 1924. Two-page list of Marconi companies and agents.

546 Fifty Years of Wireless. Chelmsford: Marconi Co., 1947 (1963). 16 pp. A brief survey to commemorate the Jubilee of the Marconi Company in 1947.

547 Hancock, Harry E., Wireless at Sea: The First Fifty Years. Chelmsford: Marconi International Marine Communication Co., 1950. 233+33 pp., 44 plates. Names, dates, extracts in the text, no other refs. List of ships mentioned, pp. xxxi-xxxiii, with page numbers.

548 Baker, W. J., A History of the Marconi Company.

London: Methuen, 1970. 413 pp., 24 plates, 28 illus. An authoritative account of the origin and progress of the company and its leading role in world-wide telecommunications and electronics development. Pt. 1, pp. 15-173, from 1897 to 1918; Pt. 2, pp. 177-297, from the end of the Great War to the late 1930s; Pt. 3, pp. 301-404, from the beginning of World War II to the mid-1960s. Major topics include transatlantic radio, the beam system, radio broadcasting, aviation and marine communications, television and radar. Full details of organization, business matters, people, events, services, products, inventions, engineering, research and development. With names, dates and text references but no separate documentation.

Metropolitan-Vickers Electrical Company

549 Rowlinson, Frank (comp.), Contributions to Victory. Manchester: Metropolitan-Vickers Electrical Co., 1947. 199 pp., illus. An account of some of the special work of the company during World War II.

Radio Corporation of America

550 "20th Anniversary." RCA Family Circle, 5:1-8, Oct. 1939. 60 small photos. Chronological entries from 1919.

551 The First 25 Years of RCA. A Quarter Century of Radio Progress. New York: RCA, 1944. 87 pp. Also, Thirty Years of Pioneering, 1949.

552 Warner, John C., and Elmer W. Engstrom, The Radio Corporation of America. Three Historical Views. New York: RCA, 1963. 20 pp., 39 illus. In 3 parts; The years to 1938, 1938-1958, 1958-1962.

Standard Telephones and Cables, Limited

553 Eve, C. W., "Standard Telephones and Cables, Limited, London--60th Anniversary." Elect. Comm., 21: 213-217, 1942-1944.

Telegraph Construction and Maintenance Company

554 The Telcon Story: 1880-1950. London: Telegraph Construction and Maintenance Co., 1950. 176 pp.

Western Electric Company

555 Lovette, Frank H., "Western Electric's First 75 Years: A Chronology." Bell Tel. Mag., 23:271-287, 1944-1945.

556 Iardella, Albert B. (ed.), Western Electric and the Bell System. A Survey of Service. New York: Western Electric Co., 1964. 115 pp. "Intended primarily for Western Electric men and women entering management positions." History of Western Electric, pp. 27-41, and other historical matter on AT&T and Bell Telephone Laboratories. App., pp. 107-114, "The Consent Decree," Jan. 1956.

557 Gorman, Paul A., Century One... A Prologue. New York: Newcomen Society, 1969. 24 pp., 15 illus. An address dealing with the history of Western Electric Company.

Western Union Telegraph Company

558 The Story of Western Union. New York: The Western Union Telegraph Co., 1958. 13 pp., 5 illus. First transcontinental line, systems, services, submarine cables, dates and some key statistics.

Westinghouse Electric Corporation

559 Woodbury, David O., Battlefronts of Industry. Westinghouse in World War II. New York: John Wiley, 1948. 342 pp., 33 plates. Additional notes, pp. 325-333, on contracts, war profits, officers and directors. A general account of the wartime activities of Westinghouse Electric Corporation with particulars of contracts, equipment and products, quantities, plants, personnel, problems, solutions and achievements. Radio, radar, X rays, instruments and power apparatus are included in this panoramic view.

Weston Electrical Instrument Corporation

560 Measuring Invisibles; the Fifty-Year Record of the World's Largest Manufacturer of Electrical Measuring Instruments. Newark, N. J.: Weston Electrical Instrument Corporation, 1938. 51 pp., illus. On Edward Weston, the growth of the company and its products, electrical apparatus and measuring instruments.

Zenith Corporation

561 The Zenith Story: A History from 1919. Chicago: Zenith Corporation, 1955.

D. Miscellaneous Company and Area Histories

The story of individual companies is, on the whole, a neglected aspect of technological history. Apart from the listings in the previous section, published information of this sort is scarce and fragmentary and scattered throughout a wide variety of journals, magazines and biographies. Many concerns have not systematically preserved the literature related to their own products and activities. Undoubtedly, much valuable material is buried in the files of successor companies, and much has been lost through fire and other hazards that affect continuity.

A variety of materials that include details of company histories exists in unpublished papers, reminiscences and biographies deposited in the archives of universities, associations, societies and museums, as well as in private collections. House organs are other likely sources of news items, short articles or summaries. Also, certain specialist magazines, such as the Antique Wireless Association's Old Timer's Bulletin (1217), contain useful historical matter on companies and their products or services.

Since little effort has been made to locate such materials for this book, the following very brief list is merely indicative of the kind of item that awaits collection and more complete documentation. A number of items in Chap. 3 and some of the biographies in Sect. 4C contain material on specific industrial organizations. See especially 267, 273, 274, 276, 287, 305, 332, 333, 336, 337, 340, 341, 396, 404, 405, 463. See also Sect. 13A, particularly 1194, 1201, 1211, 1212, and Sect. 13H.

Atwater Kent Company

562 Atlee, Frank, "Atwater Kent Manufacturing Company." The Old Timer's Bull., Vol. 7, No. 3, pp. 15-18, Autumn 1966. 3 photos, 3 diags., list of receivers.

De Forest Companies. See 305, 396, 1212.

Farnsworth Television and Radio Corporation. See 404, 1212.

Fessenden Companies. See 305, 405, 571, 1212, 1254, 1256.

A. H. Grebe and Company

563 Gray, G. Jack, "The Grebe Radio Story." The Old Timer's Bull., Vol. 3, No. 4, Winter 1962-1963. A. W. A. Monograph No. 6. 4 pp., 3 photos.

Motorola

564 Petrakis, Harry M., The Founder's Touch: The Life of Paul Galvin of Motorola. New York: McGraw-Hill, 1965. 240 pp., port., 20 plates. Much of this biography is concerned with the Galvin Manufacturing Company from 1928, later Motorola; its products, people, growth and expansion into television, semiconductors and modern electronics.

National Electric Supply Co.

565 Duvall, E., "The History of the National Electric Supply Co." The Old Timer's Bull., Vol. 2, No. 4, Autumn 1961. A. W. A. Monograph No. 2, 4 pp. (Vitro Corporation.)

New Jersey Area

566 Pierce, John R., and Arthur G. Tressler, The Research State: A History of Science in New Jersey. New York: D. Van Nostrand, 1964. Vol. 15, The New

Jersey Historical Series. 165 pp., 20 illus. Chaps. 7-10, pp. 84-125; The science of matter and motion, Radio communications, Some finer points of communication, and Radar and military electronics. Bibliographical Notes, pp. 152-155, contain a few chapter refs.

Pittsburgh Area

567 Kintner, Samuel M., "Pittsburgh's Contributions to Radio." Proc. IRE., 20:1849-1862, Dec. 1932. On the history of radio development and local events, particularly the work of Fessenden, early radiotelephony, Westinghouse Company and KDKA. No refs.

San Francisco Bay Area

568 Enochs, Hugh, "The First Fifty Years of Electronic Research." Palo Alto, Calif.: Palo Alto Chamber of Commerce. The Tall Tree, Vol. 1, No. 9, May 1958. 52 pp., illus. Account of activities by men and companies from the early years of this century. Names, dates, events and personal recollections of pioneers related to amateur and commercial radio, communication companies, manufacturing and research establishments.

569 Enochs, Hugh, Electronics Research in the Space Age. Palo Alto, Calif.: Palo Alto Chamber of Commerce, 1962. 3rd ed., 32 pp., 66 illus. The story of electronics research and development in the Palo Alto-Stanford University area. Pt. 1, pp. 3-18, company, university, and government research in more than 200 laboratories. Pt. 2, pp. 19-26, major research centers. Pt. 3, pp. 27-32, chronological history of electronic events from 1908.

570 Morgan, Jane, Electronics in the West: The First Fifty Years. Palo Alto, Calif.: National Press Books, 1967. 194 pp., 70 illus. Story of electronics research and industry in the San Francisco Bay area. Intended for young adults and classroom use, this romanticized text concentrates on men and inventions. Not documented.

Submarine Signal Company

571 Fay, H. J. W., Submarine Signal Log. Portsmouth,

R. I.: Raytheon Company, Submarine Signal Division, 1963. 37 pp., 23 illus. Underwater communication, echo sounding, the inventions of R. A. Fessenden, submarine detection, and the work of the Submarine Signal Company from 1901. See also 405.

6. BROADCASTING

The larger body of literature on broadcasting encompasses special areas such as education, religion, government, entertainment and the study of mass media, as well as production and studio techniques for radio and television. Except for a few general surveys, such works have been omitted in preference to those that are concerned more directly with the origin, development and progress of stations and systems, regulations, services, engineering and apparatus, and others that include historical and technical matter. Many items in Chap. 13 (Radio), and Chap. 16 (Television) also concern broadcasting. Parts of a few selected items in those subject chapters have been included as subentries. See also Sects. 3E, 3G.

A. General Surveys

572 Rothafel, Samuel L., and Raymond F. Yates, Broadcasting, Its New Day. New York: Century, 1925. 316 pp., illus., ports., plates, diags. This early book on broadcasting deals with programming, international and educational aspects of radio, and technical developments. Early television is mentioned.

573 Briggs, Asa, The History of Broadcasting in the United Kingdom. Vol. 1, The Birth of Broadcasting. London: Oxford University Press, 1961. 425 pp., 20 plates, 23 line illus. Bibliographical note, pp. 407-409, 29 entries. Apps. 1-3, pp. 410-415, plus 4 folding organization charts. A thorough and detailed survey from early experimental days of the late 1890s to the rapid growth of the BBC up to 1926. Fully supported with over 1050 footnote refs. Slight technical treatment. See Vol. 2, 582; Vol. 3, 630.

574 Stewart, Irwin (ed.), "Radio." Ann. Am. Acad., 142: Supp., March 1929. 107 pp. A collection of 16 articles that includes general history, international radio, U.S. legislation, with mention of amateur and military communications. A broad, nontechnical survey. See 578, 584.

575 Goldsmith, Alfred N., and Austin C. Lescarboura, This Thing Called Broadcasting. New York: Henry Holt, 1930. 362 pp., 15 plates. A popular survey: "A simple tale of an idea, an experiment, a mighty industry, a daily habit, and a basic influence in our modern civilization." A panoramic view of American broadcasting with some historical matter up to page 80. No refs., no index.

576 Barnouw, Erik, A History of Broadcasting in the United States. Vol. 1, A Tower in Babel, to 1933. New York: Oxford University Press, 1966. 344 pp., 16 plates. App. A, Chronology, 1895-1933, pp. 287-290. App. B, Laws (Radio Acts, 1912, 1927), pp. 291-315. Bibliography, pp. 317-328, has nearly 300 entries; many of unpublished items, but sources are given. Well documented with more than 650 footnotes, with mixed numbering by sections. A general, nontechnical account, with emphasis upon the personal, business, and social aspects of broadcasting. See Vol. 2, 593. Vol. 3, 642.

577 [AT&T]. Broadcasting Network Service. New York: AT&T, 1934. 53 pp. Details of the telephone system's role in network broadcasting.

578 Hettinger, Herman S. (ed.), "Radio: The Fifth Estate." Ann. Am. Acad., Vol. 177, Jan. 1935. 219 pp. 29 articles: Broadcasting systems, 11; broadcasting services, 10; current questions, 8. See 574, 584.

579 Broadcasting Yearbook. Washington: Broadcasting Magazine. Published from 1935.

580 Schrage, Wilhelm E., "Milestones in Broadcasting." Radio-Craft, 7:456, 457, 489, Feb. 1936. One photo, 4 diags., 2 tables. A review of radio broadcasting since 1920, with particular reference to current statistics.

581 Radio Annual. New York: Radio Daily Corporation. Published from 1938.

582 Briggs, Asa, The History of Broadcasting in the United Kingdom. Vol. 2, The Golden Age of Wireless. London: Oxford University Press, 1965. 663 pp., 24 plates, 19 line illus. Bibliographical note, pp. 660-663, 86 entries. Three folding charts on the organization of the BBC. From the beginning of 1927 to September 1939. Thoroughly supported by some 2250 footnotes. Slight tech-

nical treatment. Includes early mechanical television by John L. Baird and the rise of electronic television. See Vol. 1, 573; Vol. 3, 630.

583 Eckersley, Peter P., The Power Behind the Microphone. London: Jonathan Cape, 1941. 255 pp. The former chief engineer of the BBC discusses the background scene and evaluates developments of British and American broadcasting systems from 1921. Early BBC television is mentioned.

584 Hettinger, Herman S. (ed.), "New Horizons in Radio." Ann. Am. Acad., Vol. 213, Jan. 1941. 189 pp. 24 articles: Current problems in radio, 8; broadcasting as a social force today, 9; coming developments in radio, 7. Includes significance of new inventions, frequency modulation, television, facsimile. Summary of developments since 1935. See 574, 578.

585 Chase, Francis S., Sound and Fury. An Informal History of Broadcasting. New York: Harper and Brothers, 1942. 304 pp. On the medium's early days and growth; programs, personalities, and social effects. Shortwave radio, FM, and television are included.

586 Waller, Judith C., Radio, the Fifth Estate. Boston: Houghton, Mifflin, 1946. 483 pp. This comprehensive survey treats radio activities at home and abroad, stations and networks, programs, advertising, education, engineering, regulations, and so forth. There are numerous items of historical interest.

587 Rose, Oscar, Radio Broadcasting and Television: An Annotated Bibliography. New York: H.W. Wilson, 1947. 120 pp. Pt. 1, Radio broadcasting, pp. 9-89. Pt. 2, Television, pp. 93-97, 26 items. History and general survey, pp. 9-21, 86 items.

588 Coase, R.H., British Broadcasting: A Study in Monopoly. London: Longmans, Green, 1950. 206 pp. This thorough study encompasses the origins, wire broadcasting, foreign commercial broadcasting, and public discussions. Chap. 1, pp. 3-29, deals with the origins of British broadcasting, but there is much historical material in other chapters. Fully documented with 432 refs. given as notes with each chapter. Many of these are quite extensive and contain further material and refs.

589 Siepmann, Charles A., Radio, Television and Society. New York: Oxford University Press, 1950. 410 pp. Apps. (7), pp. 359-398, various legislative excerpts. Questions and selected reading matter arranged by chapter; 94 articles, books and documents. Pt. 1, Systems of broadcasting; Pt. 2, Social implications of radio and television. Chap. 1, pp. 3-14, Radio in the United States: Early history, 1920-1934.

590 Willis, Edgar E., Foundations in Broadcasting: Radio and Television. New York: Oxford University Press, 1951. 439 pp., 38 illus. Glossary, pp. 415-422. Bibliography, pp. 423-428, 96 entries in 8 groups. Brief history, pp. 3-16, otherwise primarily a textbook on broadcasting techniques.

591 Dunlap, Orrin E., Radio and Television Almanac. Many entries concern specific events in broadcasting. See 295.

592 Hanson, O. B., "Historic Highlights in Developing the Radio Broadcasting and Television Arts." Communication and Electronics, No. 3, pp. 364-371, Nov. 1952. Review of pioneering achievements from Hertz and Marconi through early radio broadcasting, frequency modulation, television. No refs.

593 Barnouw, Erik, A History of Broadcasting in the United States. Vol. 2, The Golden Web, 1933 to 1953. New York: Oxford University Press, 1968. 391 pp., 24 plates. App. A, Chronology, 1933-1953, pp. 305-309. App. B, Law (Communications Act of 1934), pp. 311-347. Bibliography, pp. 349-369, contains about 500 entries, including manuscripts and sources. Over 470 footnotes, numbered by sections. General, nontechnical, with emphasis upon personalities, programs, and the social and business aspects of broadcasting. See Vol. 1, 576; Vol. 3, 642.

594 [B. B. C.]. British Broadcasting: A Bibliography. London: British Broadcasting Corporation, 1954. 36 pp., 617 items, some multiple entries, most with brief annotations. Author index, pp. 32-35. Books, articles and official publications published in Britain, except engineering subjects.

595 Sturmey, S. G., The Economic Development of Radio. Five chapters are concerned with British broadcast-

ing: 8, pp. 137-163, Sound, 1920-1927; 9, pp. 164-189, Sound, 1927-1956; 10, pp. 190-213, Television; 11, pp. 214-235, Patents; 12, pp. 236-255, Wire relays. There are 178 refs. to articles, reports and memoranda with these chapters. See 1420.

596 Codding, George A., Broadcasting Without Barriers. Paris: UNESCO, 1959. 167 pp., 8 photos. A broad survey of international radio and television. Early history is dealt with in pp. 11-24. The body of the book is concerned with domestic systems, broadcasting in less advanced countries, international communications, use of the radio spectrum and frequency allocations, and improved techniques, including FM and wired radio. Television is treated in pp. 124-139. Distribution of radio receivers in 140 countries, pp. 147-150. Glossary of technical terms, pp. 151-153. Selected bibliography, pp. 154-158; 150 entries of books, reports, and some periodicals. Has six pages of colored pictographs showing world-wide radio audience, radio spectrum divisions and allocations, and ITU organizational structure. Many footnotes include refs.

597 Town, George R., "Frequency Allocations for Broadcasting." Proc. IRE., 50:825-829, May 1962. 16 footnote refs. from 1932. History from 1910. Includes AM, FM and television.

598 Emery, Walter B., Broadcasting and Government: Responsibilities and Regulations. East Lansing: Michigan State University Press, 1961. 482 pp. Use of radio frequencies, programming, governmental control, licensing, regulations, legal aspects. History of FM and television are treated briefly. There is a summary of regulations up to 1934, and a chronology of the FCC. Bibliography, pp. 469-473.

599 Evolution of Broadcasting. Washington: FCC, July 1961. Information Bull. 2-B, 25 pp.

600 Cole, Barry J., and Al P. Klose, "A Selected Bibliography on the History of Broadcasting." J. Broadcasting, 7:247-268, Summer, 1963. A limited selection of 149 items arranged alphabetically by authors; general and almost wholly nontechnical. Includes television.

601 Emery, Walter B., National and International Systems of Broadcasting: Their History, Operation and Control.

East Lansing: Michigan State University Press, 1969. 752 pp., 10 Apps., pp. 599-719. Bibliography, pp. 723-738, lists some 350 books and periodicals. Thoroughly documented with nearly 1300 chap. notes. Although nontechnical, this comprehensive survey furnishes useful background material and a necessary international perspective of particular value to the student primarily interested in the cultural, social, political and legal aspects of broadcasting.

B. Radio

602 History of Radio Broadcasting and KDKA. Pittsburgh: Westinghouse Electric Corporation, c. 1947. Typescript, 27 pp. Chiefly on events of the early 1920s, with a little prior history. Numerous details with names and dates, and brief mention of short waves, FM and television.

603 McLauchlin, R. J., "What the Detroit 'News' Has Done in Broadcasting." Radio Broadcast, 1:136-141, June 1922. 7 illus. History of the News broadcasting station from the first public program on August 31, 1920.

604 The Detroit News (staff), WWJ--The Detroit News. 1922. 95 pp., illus.

605 Yates, Raymond F., and Louis G. Pacent, The Complete Radio Book. There are 35 pages on radio broadcasting. See 1294.

606 Burrows, Arthur R., The Story of Broadcasting. London: Cassell, 1924. 183 pp. Eight plates include 39 ports. of BBC personalities. There is also a chart showing the organization of the BBC. The first 53 pages survey early history of radio to 1922. No index.

607 Lewis, C. A., Broadcasting from Within. London: George Newnes, 1924. 177 pp., 17 plates, mainly ports. of the BBC staff. No index.

608 Reith, John C. W., Broadcast Over Britain. London: Hodder and Stoughton, 1924. 231 pp.

609 Dowsett, H. M., Wireless Telphony and Broadcasting. Vol. 1, pp. 54-60, The initiation and early stages of broadcasting; pp. 61-66, How official British broadcasting was started. App., pp. 203-210, Post Office and BBC broad-

casting licence and agreement. Vol. 2, pp. 142-160, Broadcasting in other countries. See 1191.

610 Banning, William P., Commercial Broadcasting Pioneer. The WEAF Experiment, 1922-1926. Cambridge, Mass.: Harvard University Press, 1946. 308 pp., 31 plates, 2 tables, 6 charts. Condensed chronology (WEAF), pp. xxix-xxxiii, March 1922 to January 1927. 195 footnote refs. Much detailed background of early broadcasting. Only slightly technical.

611 Archer, Gleason L., History of Radio to 1926. About 130 pages provide valuable source material on the beginnings and early progress of radio broadcasting in the United States. Chap. 12, pp. 190-204, Westinghouse Company inaugurates radio broadcasting. Chap. 13, pp. 206-224, Pioneer days in radio broadcasting. Chap. 15, pp. 240-272, Radio's great era of expansion. Chap. 16, pp. 274-295, Radio broadcasting grows up. Chaps. 17-20, pp. 296-372, cover litigation and rivalries, the national election of 1924, the struggle for network broadcasting, wavelengths and injunctions. See 1194, also 1417.

612 Davis, H. P., "The Early History of Broadcasting in the United States." The Radio Industry, pp. 189-225. Experiments before World War I, beginnings of KDKA, public interest and support, famous events, technical developments, the radio industry, beginning of NBC and national service. Other sections of the book are also pertinent. See 1415. This article was also published as a 23-page booklet, The History of Broadcasting in the United States. Cambridge, Mass.: Harvard University Press, 1928.

613 Schubert, Paul, The Electric Word. Chap. 9, pp. 191-211, Broadcasting emerges; Chap. 10, pp. 212-249, The radio boom. See 1200.

614 Codel, Martin (ed.), Radio and Its Future. Pt. 1, pp. 3-91, Broadcasting. Eight chapters by specialists, almost wholly on American radio; history, business, advertising, education, entertainment, national and international broadcasting. Pt. 4, pp. 219-264, Regulation. See 1201.

615 Horn, C. W., "Ten Years of Broadcasting." Proc. IRE., 19:356-376, March 1931. 15 photos of transmitters and broadcasting facilities. Brief review of the beginnings of broadcasting, development of receivers, radio

tubes, transmitters, and network broadcasting. No documentation.

616 Wood, Robert S., "Twelve Years of Radio Progress." Radio News, 13:24, 25, 86, 88, 91, July 1931. 5 illus. A brief review of broadcasting.

617 Moseley, Sydney A., Broadcasting in My Time. London: Rich and Cowan, 1935. 244 pp. "A history of broadcasting in Great Britain from its earliest origins."

618 Archer, Gleason L., Big Business and Radio. This business history of the radio industry includes much on American broadcasting. Chap. 13, pp. 278-299, The National Broadcasting Company arises. Chap. 14, pp. 300-321, Travails of a rival radio network. Radio during the depression years and broadcasting through the 1930s are covered. See 1417.

619 Maclaurin, William R., Invention and Innovation in the Radio Industry. About 50 pages are devoted to radio broadcasting, particularly inventions, patents, licenses, litigation and company affairs, mainly from 1920 to 1940. See 1212.

620 "25th Anniversary of Radiobroadcasting." Elect. Eng., 64:365, 366, Oct. 1945. 4 photos. On the first established station, KDKA in Pittsburgh.

621 Settel, Irving, A Pictorial History of Radio. New York: Citadel Press, 1960. 176 pp., over 400 illus. Bibliography, p. 172, 22 items. "The Beginning," pp. 15-32, brief history with 25 illus. Highlights of developments in the United States; major events, famous shows, personalities and programs. Nontechnical.

622 Guy, Raymond F., "AM and FM Broadcasting." Proc. IRE., 50:811-817, May 1962. 2 tables, 25 refs. from 1920. Early experiments and milestones from 1906.

623 "Special Feature: Radio at 40." Broadcasting, May 14, 1962. A special historical report in 5 sections. The future after 40-plus, pp. 77-80, 8 photos. First great year of growth, pp. 82, 84, 88, 90, 92, 96, 98, 100, 102-104, 106, 110, 112, 114, 116, 118-120, 122, 40 photos, many of early equipment. Pioneers still in business, pp. 122-137, list of

141 stations, 16 photos. Days of hand-made hardware, p. 138, 2 photos. Accounts that took a chance, pp. 139, 140, 2 photos.

624 BBC Sound Broadcasting: Its Engineering Development. London: British Broadcasting Corporation, 1962. 96 pp., 54 photos, 5 maps. Important dates in the development of BBC sound broadcasting, pp. 92-96, 205 entries with specific dates from Nov. 14, 1922 to June 27, 1962. The beginning of broadcasting in the United Kingdom up to 1939, pp. 5-17. Wartime developments, pp. 17-27. The postwar period, pp. 28-56. External broadcasting, pp. 56-77. This monograph was published to mark the 40th anniversary of the BBC. Also, Pawley, Edward L. E., "B. B. C. Sound Broadcasting 1939-60." Proc. IEE., 108 (Pt. B):279-302, May 1961.

C. Radio and Television

625 Waldrop, Frank C., and Joseph Borkin, Television: A Struggle for Power. New York: William Morrow, 1938. 299 pp. An economic and social study on the problems of, and the possibilities for, the new medium in the United States, especially in regard to business interests versus public interest. Includes material on radio and the larger companies, competitive activities, general stream of development, technical progress, regulations, patent rights, radio frequencies and spectrum uses, and censorship.

626 Thomas, Lowell, Magic Dials. The Story of Radio and Television. New York: Lee Furman, 1939. 142 pp. Largely a pictorial essay (with many good photographs, some in color) on American broadcasting. Major historical events are highlighted. Popular treatment.

627 Lohr, Lenox R., Television Broadcasting: Production, Economics, Technique. New York: McGraw-Hill, 1940. 274 pp., 86 illus. This early text on the title topics includes some general history and technicalities with details of NBC and RCA. No refs.

628 Porterfield, John, and Kay Reynolds (eds.), We Present Television. New York: W. W. Norton, 1940. 298 pp., 16 plates, 4 diags. List of television broadcast stations as of May 1, 1940, pp. 283-285, 23 entries. Biographical notes, pp. 287-293, 14 entries. Glossary of terms,

pp. 295-298. An informal record of the new industry for the layman by 12 contributors. Some historical background, little technical matter, no refs., but some dates in the text.

629 Hubbell, Richard W., _4000 Years of Television._ American, British and foreign broadcasting companies and systems are mentioned. _See_ 1646.

630 Briggs, Asa, _The History of Broadcasting in the United Kingdom._ Vol. 3, _The War of Words._ London: Oxford University Press, 1970. 766 pp., 24 plates, 8 other illus. Bibliographical note, pp. 727-732, 95 entries. Thorough coverage of the role of British radio services from 1939 to 1945, fully supported with about 2500 footnotes. _See_ Vol. 1, 573; Vol. 2, 582.

631 Tyler, Kingdon S., _Telecasting and Color._ New York: Harcourt, Brace, 1946. 213 pp., 41 full-page illus., 11 figures in the text, colored frontispiece of the CBS color system for film. List of books on television, pp. 202-204, 39 entries. A general introduction to television broadcasting and technicalities for the layman. Color television treated in pp. 145-179, primarily for the CBS system, with a little on J. L. Baird's method.

632 Swift, John, _Adventure in Vision. The First Twenty-Five Years of Television._ London: John Lehmann, 1950. 223 pp., 32 plates, folding studio plan, 6 diags. An over-all story for the layman, mainly about British progress--John Baird, the BBC, people, programs and activities--with mention of developments in France, Germany and the United States.

633 Fink, Donald G., "Television Broadcasting in the United States, 1927-1950." _Proc. IRE._, 39:116-123, Feb. 1951. An overview that includes standards, program service, specifications, channel frequencies, theater television, color television. No refs. Also, "Television Broadcasting Practice in America: 1927-1944." _J. IEE._, 92 (Pt. 3):145-164, Sept. 1945.

634 Gorham, Maurice A. C., _Broadcasting and Television Since 1900._ London: Andrew Dakers, 1952. 274 pp. Primarily a history of early radio and of the British Broadcasting Corporation. General and nontechnical, with mention of developments in the United States and other countries. Various footnotes, no refs.

635 Abramson, Albert, Electronic Motion Pictures.
Covers various aspects of television broadcasting, with numerous refs. See 1661.

636 Head, Sydney W., Broadcasting in America; a Survey of Television and Radio. Boston: Houghton, Mifflin, 1956. 502 pp., illus. This broad study is concerned with origins and growth, basic technical elements, economic and social aspects, controls and regulations.

637 Paulu, Burton, British Broadcasting: Radio and Television in the United Kingdom. Minneapolis: University of Minnesota Press, 1956; London: Oxford University Press, 1957. 409 pp. Comprehensive survey and evaluation of British broadcasting; programming, operations, facilities, structure, finance and development. Also deals with the Independent (I. T. A.) system.

638 Crozier, Mary, Broadcasting: Sound and Television. London: Oxford University Press, 1958. 236 pp. On the development of major broadcasting systems and sociological effects, with historical material. Includes a bibliography of radio and television broadcasting.

639 Ross, Gordon, Television Jubilee. The Story of 25 Years of BBC Television. London: W. H. Allen, 1961. 224 pp., 33 plates. A popular treatment of events and personalities. No index.

640 Owen, Clure H., "Television Broadcasting." Proc. IRE., 50:818-824, May 1962. 7 illus. Partly historical. 14 refs. from 1927.

641 Dizard, Wilson P., Television. A World View. Syracuse, N. Y.: Syracuse University Press, 1966. 349 pp., 16 plates. App., pp. 293-297, a statistical survey of overseas television. Notes to chapters, pp. 299-316, contain 349 entries. Selected bibliography, pp. 321-333, has some 270 entries. A sound survey of the growth of international television during recent years, but with slight technical interest.

642 Barnouw, Erik, A History of Broadcasting in the United States. Vol. 3, The Image Empire, from 1953. New York: Oxford University Press, 1970. 396 pp., 20 plates. App. A, Chronology, 1953-1970, pp. 345-348. App. B, Laws, pp. 349-354. Bibliography, pp. 355-373, con-

tains about 470 entries, including manuscripts and sources. About 480 footnotes, numbered by sections. See Vol. 1, 576; Vol. 2, 593.

7. ELECTRICITY AND ELECTRONICS

The earlier books and articles collected in this chapter portray the beginnings and progress of electronics, particularly the core subject--conduction of electricity in rarefied gases. Histories, surveys and textbooks comprise the major part of the list, supplemented by monographs and leading papers. The books were selected as being most representative of contemporary practices; for their record of theories and experiments and for the illustrations of apparatus and techniques.

A number of the most voluminous works of past times also contain extracts from original sources and comments upon current thought and activities, frequently with miscellaneous references. These features are excellent supplements to the more orthodox historical accounts. Also, many 19th-century authors devoted portions of their text to historical surveys of the different branches of electrical science, a practice that has greatly declined in recent decades. This earlier attention to history is fortunate because, even though many references are to secondary sources, minor works and events of passing interest, they provide excellent guides to material that is largely forgotten, some probably not otherwise recorded, and much that is often difficult to trace. See also Chap. 3, especially 189, 190, 200, 236, 238, 244, 252.

A. Electricity and Magnetism

643 Martin, Thomas (ed.), Faraday's Diary. London: G. Bell and Sons, 1932-1936. 8 vols., including index vol. Essentially laboratory notes covering the period from 1820 to 1862. A detailed record of experiments, apparatus, effects, results and related suppositions and conclusions, supported by extensive sketches. See Nature, 126:812-814, Nov. 22, 1930, also 712. See also Michael Faraday's Experimental Researches in Electricity, 1839-1855, London: R. and J.E. Taylor, 3 vols.

644 Noad, Henry M., A Manual of Electricity. London:

George Knight, 1855, 1857. 4th ed., Lockwood, 1859. 910 pp., 497 wood engrs. Early history of statical electricity, pp. 1-16; historical sketches in some other chapters and much historical matter throughout. There is an index to authorities, pp. 903-910, in addition to a subject index. Numerous short references to sources are given in the text. A few pages (738-742) deal with conduction of electricity through gases. The text, highly descriptive and nonmathematical, is an interesting and informative record of experimental techniques. There are many paraphrased accounts and extracts from original sources. Often quoted as an authority by writers up to 1900, this voluminous work amply portrays the "state of the art" at mid-century. One chapter describes the electric telegraph, see 1508.

645 Maxwell, James C., A Treatise on Electricity and Magnetism. Oxford: Clarendon Press, 1873. 2 vols., 20 plates. 2nd ed., 1881, with preface by W. D. Niven. 3rd ed., 1892. 506, 500 pp., 39, 68 illus., numerous tables and many footnotes, some refs. Preface by J. J. Thomson. Reprinted various years, also New York: Dover, 1954. One of the great books of electrical science.

646 Guthrie, Frederick, Magnetism and Electricity. London and Glasgow: W. Collins; New York: G. P. Putnam's Sons, 1876. 364 pp., 301 engrs. Frictional or static electricity, voltaic electricity, magnetism. Apps., pp. 351-359, laboratory hints and instructions for making simple electrical apparatus.

647 Gordon, James E. H., A Physical Treatise on Electricity and Magnetism. London: Sampson, Low; New York: D. Appleton, 1880. 2 vols. 323 pp., 26 plates, 148 diags.; 295 pp., 26 plates, 256 diags. Index in Vol. 2. Vol. 1, electrostatics, magnetism, electro-kinetics; Vol. 2, electro-kinetics, electro-optics. Treats electrical science from the physical viewpoint. There are many illustrations of instruments and equipment, with full details of, and liberal quotations from, original experiments and papers. Numerous footnotes give references to books and papers.

648 Maxwell, James C. (William Garnett, ed.), An Elementary Treatise on Electricity. Oxford: Clarendon Press, 1881. 208 pp., 6 plates, 53 diags. This text was compiled from the original incompleted manuscript with selected portions taken from the author's Electrcity and Magnetism. 2nd ed., 1888. See 645.

649 Thompson, Silvanus P., Elementary Lessons in Electricity and Magnetism. London: Macmillan, 1881, 1884, 1886. 456 pp., 171 illus. A thorough introduction for beginners centered on theories, laws, experiments and apparatus. Chap. 12, pp. 401-420, deals briefly with telegraphs and telephones. Historical matters are mentioned throughout. This basic text was continually reprinted and revised; 7th ed., 1924, 706 pp., 377 illus.

650 Lodge, Oliver J., Modern Views of Electricity. London: Macmillan, 1889. 422 pp., illus. 2nd ed., 1892, 3rd ed., 1907. 518 pp., 67 illus. On electrostatics, conduction, magnetism, radiation and the ether, with 6 appended lectures, pp. 345-477. App. (mathematical notes), pp. 481-502. About 90 footnotes, some refs., other refs. in the text.

651 Thomson, Joseph J., Notes on Recent Researches in Electricity and Magnetism, Intended as a Sequel to Professor Clerk Maxwell's Treatise on Electricity and Magnetism. Oxford: Clarendon Press, 1893. 578 pp., 144 illus. A full survey of advanced electrical theories. Reprinted 1968. See also 722.

652 Thomson, Joseph J., Elements of the Mathematical Theory of Electricity and Magnetism. Cambridge: University Press, 1895. 510 pp., 132 illus. 2nd ed., 1897, 508 pp. 3rd ed., 1904, 544 pp. 4th ed., 1909, 550 pp., with additions and corrections. 5th ed., 1921, 420 pp.

653 Trowbridge, John, What is Electricity? New York: D. Appleton, 1896. 315 pp., 55 illus. A popular survey of current scientific views with special reference to energy, radiations, and the ether. Chaps. 17-20, pp. 215-297, are concerned with wave motion, electric waves, the electromagnetic theory of light and the ether, and X rays.

654 Thomson, Joseph J., Electricity and Matter. London: Constable; New York: Charles Scribner's Sons, 1904. 162 pp., 18 illus. Exposition on recent advances in electrical science, from the Silliman Memorial lectures at Yale University, May 1903. No index.

655 Fleming, John A., "Electricity." Encyclopaedia Britannica, 11th ed., 9:179-193. 27 footnote refs. Historical survey of theories and applications. Electronic theory, pp. 192, 193. Bibliography, 24 items.

656 Fleming, John A., "Electromagnetism." Encyclopaedia Britannica, 1911. 11th ed., 9:226-232. 6 diags. and graphs, 7 tables. 3 footnotes, bibliography, 13 items.

657 Bidwell, Shelford, "Magnetism." Encyclopaedia Britannica, 1911. 11th ed., 17:321-353. 29 diags. and graphs, 28 tables and lists. A full survey of theory, principles, experiments and effects in 13 sections, from general phenomena to the molecular theory of magnetism. 93 footnotes, mainly refs., other refs. in the text. Historical and chronological notes, pp. 351-353, 13 footnote refs. and 15 items in a reading list.

658 Glazebrook, Richard T. (ed.), A Dictionary of Applied Physics, Vol. 2, Electricity. Articles on "Electricity," and "Magnetism." See 51.

659 Whittaker, Edmund T., A History of the Theories of Aether and Electricity. London: Thomas Nelson, 1951, 1953. New York: Philosophical Library, 1951; Harper, 1961. Vol. 1, The Classical Theories, 434 pp. From early times to 1900. Name index only. Vol. 2, The Modern Theories, 1900-1926, 319 pp. Includes subject index for this vol. only. This erudite work is heavily documented with more than 2000 footnotes that give references and supplementary sources. Vol. 2, pp. 239-252, Magnetism and electromagnetism, 1900-1926. Vol. 1 was first published in London by Longmans, Green, 1910.

660 Loeb, Leonard B., Fundamentals of Electricity and Magnetism. New York: John Wiley, 1931. 2nd ed., 1938, 554 pp. Chap. 1 (47 pp.), Historical, gives an overview of the general development of physical science and summarizes the development of electricity and magnetism over eight periods from 1600 to the mid-1930s.

661 Bates, Lesley F., Modern Magnetism. Cambridge: University Press, 1939. 339 pp., illus., diags. 2nd ed., 1948, 440 pp. 3rd ed., 1951, 506 pp. 4th ed., 1961, 514 pp., 132 diags., 29 tables. 648 footnote refs.

662 Bozorth, Richard M., "Magnetism." Rev. Mod. Phys., 19:29-86, Jan. 1947. 82 diags. and graphs, 5 tables. Theory, experiments, methods and materials, with historical matter. 11 footnote refs., bibliography, 17 items. Also, Ferromagnetism. New York: D. Van Nostrand, 1951. 968 pp., 825 illus. Bibliography, pp. 875-946, 1780 entries

from 1842. Basic text with much useful historical matter. About 650 footnote refs.

663 Skilling, Hugh H., Exploring Electricity. Man's Unfinished Quest. New York: Ronald Press, 1948. 277 pp., 20 illus. Historical survey with biographical background and several extracts in the text. List of brief biographies, pp. 253-272, 110 entries. No refs.

664 Lee, Eric W., Magnetism. Harmondsworth, Middlesex; Baltimore: Penguin Books, 1963. 287 pp., 32 plates, 61 diags. and graphs. Bibliography, pp. 274-276. History is treated in pp. 7-48, and included elsewhere in the text. Reprinted, New York: Dover, 1970.

B. Electrons

The general history of the discovery of the electron in 1897 is extensively documented in books on physics and others on general science. Highlights of progress centered on this event are found also in related works more directly concerned with X rays, cathode rays and electricity in gases, photoelectricity, atomic theory, radioactivity, nuclear physics, and quantum mechanics. Most of the entries in Sect. 3A mention historical events connected with the rise of the electron theory. See 192, 197, 200, 208, 212, for general surveys and extracts from original papers, also 213 for the respective Nobel lectures. See also Sect. 7D, particularly 722, 724, 726-728, 733.

665 Romer, Alfred, "The Speculative History of Atomic Charges, 1873-1895." Isis, 33:671-683, June 1942. 48 footnote refs. from 1834. See 670.

666 Thomson, Joseph J., "Cathode Rays." Proc. Roy. Inst., 15:419-432, April 1897. Reprinted, Smithsonian Institution, Annual Report for 1897, pp. 157-168 (Washington, 1898). 10 illus. on 3 plates. Description of apparatus and experiments used in determining the ratio of charge to mass (e/m) of corpuscles comprising cathode rays, with brief history from the work of J. Plücker in 1859. See next entry.

667 Thomson, Joseph J., "Cathode Rays." Phil. Mag., 44(Ser. 5):293-316, Oct. 1897. 6 illus. Further ac-

count of experiments (see previous entry) which identified the nature of electrified corpuscles, or electrons. Charge carried by cathode rays, electrostatic and magnetic deflection, conductivity of a gas, velocity of cathode rays, electrode materials. See also 218, pp. 77-100.

668 Kaufmann, W., "The Development of the Electron Idea." Electrician, 48:95-97, Nov. 9, 1901. Also editorial, p. 94. A survey of theories on the nature of electricity.

669 Fleming, John A., "The Electronic Theory of Electricity." Proc. Roy. Inst., 17:163, 1902. Pop. Sci. Mon., 61:5-23, May 1902. London: The "Electrician" Publishing Co., 1902. Early popular work on electronic theory and the first use of the term in the literature.

670 Romer, Alfred, "The Experimental History of Atomic Charges, 1895-1903." Isis, 34:150-161, Autumn 1942. 5 illus. 99 footnote refs. from 1881. See 665.

671 Thomson, Joseph J., Conduction of Electricity Through Gases. Cambridge: University Press, 1903. 3rd ed., J. J. and G. P. Thomson, 1928. Vol. 1, especially Chap. 6, pp. 229-290, Determination of the ratio of the charge to the mass of an ion; Chap. 7, pp. 291-309, Determination of the charge carried by the negative ion. See 728.

672 Lodge, Oliver J., Electrons, or the Nature and Properties of Negative Electricity. London: George Bell, 1906. 230 pp., 24 illus. 49 footnotes, some refs., other refs. in the text. Historical background, pp. 19-30, foreshadowing of the atom, the electron, and cathode rays. An early text on the subject, based on lectures given at the Institution of Electrical Engineers.

673 Thomson, Joseph J., "Carriers of Negative Electricity." Nobel lecture, Dec. 11, 1906. Presentation, lecture, biography. See 213, pp. 141-155.

674 Thomson, Joseph J., The Corpuscular Theory of Matter. London: Constable, 1907. 172 pp. On the electrical theory of matter and an account of the author's experiments.

675 d'Albe, E. E. Fournier, The Electron Theory. A Popular Introduction to the New Theory of Electricity and Magnetism. London: Longmans, Green, 1907. 2nd

ed., 1908. 3rd ed., 1909. 327 pp., 35 diags., port., G. Johnstone Stoney. Preface by G. J. Stoney, pp. v-xx. App., pp. 305-319, Recent progress. Chap. 2, pp. 7-22, Origin and development of the electron theory. Refs., pp. 321-323, 46 items from 1879.

676 Cox, John, Beyond the Atom. Cambridge: University Press, 1913. 151 pp., 1 plate, 11 diags. and graphs. Bibliography, pp. 147, 148, 18 items. On the discoveries of the last decade which "have led beyond the atom."

677 Millikan, Robert A., The Electron: Its Isolation and Measurement and the Determination of Some of its Properties. Chicago: University Press, 1917. 268 pp., 33 illus., 14 tables. Historical matter (pp. 6-63), with descriptions of the author's pioneer experiments. 169 footnote refs.

678 Crowther, James A., Ions, Electrons, and Ionizing Radiations. London; New York: Longmans, Green, 1919. 5th ed., 1929. 353 pp., 112 diags., 2 plates, 16 tables. Gaseous conduction, ions, the discharge tube, cathode rays, positive rays, emission by hot bodies, photoelectricity, X rays, other rays and radiations, the electron theory. Numerous historical refs. in the text. 126 chap. refs.

679 Crowther, James A., "Electrons and the Discharge Tube." Dictionary of Applied Physics, Vol. 2, pp. 337-361. 12 diags. Survey of electron theory up to 1921, with details of experiments, apparatus and results. 76 footnote refs., no titles. See 51.

680 Lodge, Oliver J., Atoms and Rays. An Introduction to Modern Views on Atomic Structure and Radiation. London: Ernest Benn, 1924. 208 pp.

681 Compton, Arthur H., X-Rays and Electrons. New York: D. Van Nostrand, 1926. 403 pp., 130 illus., 35 tables. 396 footnotes, mainly refs., some multiple entries. A theoretical work with numerous historical references.

682 Thomson, Joseph J., Beyond the Electron. Cambridge: University Press, 1928. 43 pp. Lecture at Girton College, March 3, 1928. A popular treatment of

wave mechanics, particularly waves of the free electron.

683 Thomson, George P., The Atom. London: Thornton Butterworth; New York: Henry Holt, 1930. 252 pp., 15 diags. Bibliography, pp. 245-247. Folding chart showing periodic classification of the elements. The author has "devoted a fair amount of space to the new wave theory of mechanics and atomic structure."

684 Millikan, Robert A., Electrons (+ and -), Protons, Photons, Neutrons, and Cosmic Rays. Chicago: University Press, 1935. 492 pp., 98 illus., 14 tables. Scholarly discussion with historical treatment by a pioneer investigator. 350 footnote refs. The early history of electrical theory and related experiments is treated in pp. 6-44.

685 Compton, Karl T., "The Electron; Its Intellectual and Social Significance." Science, 85:27-37, Jan. 8, 1937; Nature, 139:229-240, Feb. 6, 1937. Background, discovery, quantum theory, structure of the atom, the electron in industry.

686 Davisson, Clinton J., "The Discovery of Electron Waves." B.S.T.J., 17:475-482, July 1938. 4 illus. Nobel lecture, Dec. 13, 1937. Also 213, pp. 387-394; biography, pp. 395, 396. See next entry.

687 Thomson, George P., "Electronic Waves." Nobel lecture, June 7, 1938. See 213, pp. 397-403. Biography, pp. 404, 405. Presentation speech (including previous entry), pp. 381-386.

688 Stranathan, James D., The "Particles" of Modern Physics. Philadelphia: Blakiston, 1942. 571 pp., 217 diags. and photos, 41 tables. Chaps. 1 to 4, pp. 1-146, and Chap. 6, pp. 212-259, are particularly pertinent: Gaseous ions, the electron, electrical discharge, cathode rays, the photoelectric effect. Technical, with strong historical treatment. Extensively documented with 1128 footnote refs., titles not given.

689 Crowther, James A., "The Electron." Electronic Engineering, 19:343-347, Nov. 1947. On events leading to the acceptance of the corpuscular theory in preference to the wave theory of cathode rays. Summary of a lecture in celebration of the Electron Jubilee.

690 Weintroub, S., "Jubilee of the Discovery of the Electron." Nature, 160:776-778, Dec. 6, 1947. Summary of several lectures given during the celebrations in London.

691 Anderson, David L., The Discovery of the Electron. The Development of the Atomic Concept of Electricity. Princeton, N.J.: D. Van Nostrand, 1964. 138 pp., 10 diags., 4 tables. Traces the growth of theories and ideas with a light treatment of apparatus and techniques. Background chronology, pp. 2-5, from antiquity to 1831. A clear and concise study with 125 chap. refs. Bibliography, p. 135, 9 items.

C. X Rays

The discovery of X rays in 1895 and the discovery of the electron two years later were the fruits of some 40 years of research with electricity in gases. These epochal events became the foundation of 20th-century physics and clearly separate modern concepts from the older classical theories of electricity and matter. The X-ray story, like the history of electron theory (Sect. 7B), is embodied in numerous physics texts and books on general science, as well as in works on the history of physics. See Sect. 3A; 192, 197, 198, 200, 204, 205, 212, 214, for surveys and extracts. See also 724 for events before 1895.

692 Glasser, Otto, Wilhelm Conrad Röntgen and the Early History of the Roentgen Rays. This definitive work lists 49 books and pamphlets on Roentgen rays published during 1896, 13 of which are in English. Four of these are listed below following Röntgen's papers. There is also an extensive list of papers on Roentgen rays published in 1896. The 964 entries give all titles in English; these include about 420 items published in British and American journals, many of medical and photographic interest. See 449.

693 Röntgen, Wilhelm K., "On a New Kind of Rays."
Originally published in two parts, the first as a "preliminary communication," in the Annals of the Physical Medical Society of Würzburg, Dec. 1895, March 1896. Pt. 1, Nature, 53:274, Jan. 23, 1896; Electrician, 36:415-417, Jan. 24, 1896; Science, 3:227-231, Feb. 14, 1896. Pt. 2, Electrician, 36:850, 851, April 24, 1896; Science, 3:726-729, May 15, 1896. See 449, pp. 16-28, 216-221, for a full translation,

also 192, pp. 600-610, and 711 for extracts. See also, Smithsonian Institution, Annual Report for 1897, pp. 137-155 (Washington, 1898); Sci. Am., 74:51, 67, 82, 115, Jan. 25 to Feb. 22, 1896.

694 Meadowcroft, William H., The ABC of the X Rays. London: Simpkin; New York: American Technical Book Co., 1896. 189 pp., frontispiece, plates, diags.

695 Thompson, Edward P., Roentgen Rays and Phenomena of the Anode and Cathode. Principles, Applications and Theories. New York: D. Van Nostrand; London: E. and F. N. Spon, 1896. 190 pp., port., diags.

696 Thornton, Arthur, The X Rays. London: Percy Lund, 1896. 63 pp. Popular survey of X rays and photography.

697 Thompson, Silvanus P., Light; Visible and Invisible. London: Macmillan, 1897. 294 pp., illus. A series of lectures at the Royal Institution of Great Britain, Christmas, 1896. An early popular account of the nature of light with one chapter on X rays.

698 Hyndman, Hugh H. F., Radiation. Pages 206-274 discuss Röntgen rays. See 876.

699 Barker, George F. (trans., ed.), Röntgen Rays; Memoirs by Röntgen, Stokes, and J. J. Thomson. New York: Harper and Brothers, 1899. 75 pp. Bibliography, p. 74.

700 Jauncey, G. E. M., "The Birth and Early Infancy of X-Rays." Am. J. Phys., 13:362-379, Dec. 1945. Primarily about Röntgen's discovery, with some early history of cathode rays from 1859. Not documented, although some names, dates and extracts are given.

701 Pullin, V. E., and W. J. Wiltshire, X Rays Past and Present. London: Ernest Benn, 1927. 229 pp., 43 illus., including 21 plates. Beginnings of radiology, pp. 15-24, from Hauksbee's experiments (1705) to the work of Crookes (1879). Röntgen's discovery, pp. 25-33. Early X ray history, pp. 49-59. Early X ray apparatus, pp. 128-139. Early applications of X rays, pp. 140-148. Other parts discuss the electron, atomic structure, energy, crystals, X rays in medicine and surgery and in industry, and modern apparatus. No refs.

702 Sarton, George, "The Discovery of X Rays." Isis, 26:349-363, March 1937. A clear and succinct survey of prime events from the observations of Jean Picard (1675) to the discoveries of the late 1890s. Numerous extracts and full documentation in the text and in 33 footnotes.

703 "Fifty Years of X Rays." Engineering, 160:380, 381, Nov. 9, 1945.

704 Crowther, James A., "The Discovery of X Rays." Electronic Engineering, 17:755-759, 1945.

705 "Röntgen Commemoration. Fiftieth Anniversary of the Discovery of X Rays." Electrician, 135:527-531, Nov. 16, 1945. Report on the historical conference at the Royal Society, historical exhibit of the British Institute of Radiology, talk by Sir Lawrence Bragg, I. E. E. tributes to Röntgen, with notices on crystal analysis and industrial radiology.

706 Coolidge, William D., "Some Experiences With the X Ray." Elect. Eng., 64:423-426, Dec. 1945. Photo of Röntgen. No refs.

707 Coolidge, William D., and E. E. Charlton, "X Ray History and Development." Elect. Eng., 64:427-432, Dec. 1945. 6 illus., 2 tables, 34 refs. from 1896.

708 "X Rays an Early Institute Topic." Elect. Eng., 64:435, 436, Dec. 1945. Excerpts from papers presented at meetings a half century ago.

709 van der Tuuk, J. H., "50 Years of X Ray Progress in Europe." Elect. Eng., 64:444-448, Dec. 1945. 7 photos. Aspects of development, early improvements and applications, with mention of some historical events.

710 Bleich, Alan R., The Story of X Rays from Röntgen to Isotopes. London: Constable; New York: Dover, 1960. 186 pp., frontispiece and 54 illus. Glossary, pp. 165-175. Reading list, pp. 177, 178, 22 books and articles. Slight historical interest.

711 Dibner, Bern, The New Rays of Professor Röntgen. Norwalk, Conn.: Burndy Library, 1963. 56 pp., 35 illus. Reproduction of an article on Röntgen's announcement in The Electrician, January 24, 1896, pp. 52-54. Bibliography, p. 55, 28 entries. 43 footnotes, mainly refs. Brief

survey of electricity in gases, cathode-ray tubes, early experiments, Röntgen and his work, his discovery of the unknown rays, contemporary events and subsequent applications.

D. Cathode Rays, Tubes and Oscilloscopes

The beginnings of serious research on cathode rays is recorded in the papers of Michael Faraday and others in England and in those of Julius Plücker, Wilhelm Hittorf, Eugen Goldstein and others in Germany and France from the late 1850s. By 1900, these researches had produced a clearer understanding of electrical discharges in gases and of the particulate nature of cathode emissions, and resulted in the first practical cathode-ray indicator tube introduced by Karl Ferdinand Braun in 1897 (see 389). Between 1900 and 1930, the low-voltage cathode-ray tube and associated circuits slowly developed into practical oscilloscopes and also became the heart of electronic television receivers (Chap. 16). The rise of electron optics from the mid-1920s and further tube developments led to the electron microscope (Sect. 7F), adaptation of the cathode-ray tube as a radar indicator (Chap. 12), and a host of beam deflection devices. See also. Sect. 7B, particularly 666, 667, 678, 679, 688, 691; Sect. 7C, 700, 702, 711.

712 Faraday, Michael, Diary. Vol. 7, pp. 412-461, on the discharge of electricity in rarefied gases. Faraday's experiments and those he witnessed at Mr. Gassiot's home are described in detail, with more than 80 reproductions of sketches showing discharge tubes, striated patterns and other features of the apparatus. Entries from Jan. 23, 1858. See 643.

713 Plücker, Julius, "On the Action of the Magnet upon the Electrical Discharge in Rarefied Gases." Phil. Mag., 16(Ser. 4):119-135, Aug. 1858; 408-418, Dec. 1858; 18:1-20, July 1859. One plate with 14 diags. of tubes and apparatus and 7 patterns of the striated discharge, 2 other diags. Trans. by F. Guthrie from Ann. der Phys., 103:88; 104:113, 622, 1858.

714 Plücker, Julius, "Abstract of a Series of Papers and Notes concerning the Electric Discharge Through Rarefied Gases and Vapours." Proc. Roy. Soc., 10:256-269, 1859-1860. A survey of Plücker's papers originally published in German.

715 Gassiot, John P., "On the Stratifications and Dark Band in Electricial Discharges as observed in Torricellian Vacua." Phil. Trans., 148:1-16, 1858; 149:137-160, 1859. 12 illus., including 1 plate showing tubes, apparatus,

and striated discharges; 15 other diags. 9+15 footnotes.

716 de la Rive, August A., "Researches on the Phenomena which Characterize and Accompany the Propagation of Electricity in Highly Rarefied Elastic Fluids." Smithsonian Institution, Annual Report for 1863, pp. 169-192 (Washington, 1864). Conduction of electricity in gases, general phenomena, stratifications, influence of magnetism, with details of experiments and results. Trans. from the French.

717 Hittorf, Johann W., "On the Conduction of Electricity by Gases." Original in German, Ann. der Phys., Vol. 136, 1869. For an extract, see 192, pp. 561-563. See also 362, pp. 344-350.

718 de la Rue, Warren, and H. W. Müller, "Experimental Researches on the Electric Discharge with the Chloride of Silver Battery. The Discharge in Exhausted Tubes." Phil. Trans., 169:155-233, 1878. Also 171:65-114, 1880; 174:477-515, 1883. These papers contain a number of outstanding early photographs of the striated discharge.

719 Crookes, William, "On the Illumination of Lines of Molecular Pressure, and the Trajectory of Molecules." Phil. Trans., 170:135-164, 1879; Proc. Roy. Soc., 28:103-111, 1879. An extract is given in 192, pp. 564-576. See next entry, and 391. See also 362, pp. 344-350.

720 Crookes, William, "On Radiant Matter." Am. J. Sci., 18(Ser. 3):241-262, Oct. 1879. A British Association lecture at Sheffield, Aug. 22, 1879. 15 photos of the various tubes employed in demonstrating the properties of cathode rays, or "radiant matter." A famous paper. Also, Nature, 20:419-423, 436-440, 1879. See also, "On a Fourth State of Matter." Proc. Roy. Soc., 30:469-472, 1880; Nature, 22:153, 154, 1880. For extracts, see 391, and 647, pp. 112-130.

721 Gordon, James E. H., A Physical Treatise on Electricity and Magnetism. Vol. 2, Chap. 35, pp. 67-87, on the striated discharge. 10 plates and 7 diags., showing tube construction, apparatus and photos of the discharge patterns. 16 footnotes, mainly refs. to original papers. See 647.

722 Thomson, Joseph J., Notes on Recent Researches in Electricity and Magnetism. Chap. 2, pp. 53-207,

Passage of electricity through gases, is the first comprehensive account in an English text. Includes much material, with many refs., on the prior work of others on the conduction of electricity in rarefied gases, including thermionic emission. See 651.

723 Perrin, Jean B., "The Negative Charges in the Cathode Discharge." A Source Book in Physics, pp. 580-583. Original in French, Compt. Rend., 121:1130-1134, 1895. See 192.

724 Müller, Alex, "The Background of Röntgen's Discovery." Nature, 157:119-121, Feb. 2, 1946. A concise account of scientific research on the passage of electricity through rarefied gases up to 1895. 23 refs. from 1836.

725 Fleming, John A., "An Experiment Showing the Deflection of Cathode Rays by a Magnetic Field." Electrician, 38:302, Jan. 1, 1897. 4 diags. Brief mention of Crookes shadow tube with axial coil showing distortion of the image (magnetic focusing).

726 Harvey, E. Newton, "Radioluminescence." A History of Luminescence, Chap. 12, pp. 410-422. Electrical discharges in vacua from 1858, cathode rays, anode rays, X rays, rays from radium, scintillation counters and new particles. Also, Electroluminescence (up to 1900), pp. 296-304. See 194.

727 Whittaker, Edmund T., "Conduction in Solutions and Gases from Faraday to the Discovery of the Electron." A History of the Theories of Aether and Electricity, Vol. 1, Chap. 11, pp. 335-366. Faraday's observations in 1838 (p. 349), cathode-ray theories and experiments, photoelectricity, X rays, J. J. Thomson's researches and the ratio m/e, canal rays, ionic charge, electrons, ionization. There are 71 footnotes, many multiple entries, mostly refs. to sources. See 659.

728 Thomson, Joseph J., Conduction of Electricity Through Gases. Cambridge: University Press, 1903. 566 pp., illus. 2nd ed., 1906, 678 pp. 3rd ed., by J. J. Thomson and George P. Thomson, 2 vols., 1928, 1933. Reprinted, New York: Dover, 1969. Vol. 1, General properties of ions, ionization by heat and light. 491 pp., 2 plates, 121 diags. and graphs, numerous tables. Vol. 2, Ionization by collision and the gaseous discharge. 608 pp., 5 plates, 232

diags. and graphs, numerous tables. Fully documented with about 1500 footnotes, mainly refs. to original papers, many with multiple entries. Titles not given. The foundation work in English on cathode rays and related phenomena. See 671.

729 Lenard, Philipp E. A., "On Cathode Rays." Nobel lecture, May 28, 1906. See 213, pp. 105-130. 11 photos and diags., 25 footnotes. Presentation speech, pp. 101-104. Chronological list of publications, pp. 131-134, 55 entries, some multiple items, from 1860. Biography, pp. 135-138.

730 Thomson, Joseph J., "Electric Conduction Through Gases." Encyclopaedia Britannica, 1911. 11th ed., 6:864-890. 23 diags. and graphs, 15 tables. Gases, mixtures and pressures, electrodes, ions, charges, mass and velocity, ionization, striations, effects of ultra-violet light and X rays, corpuscles (electrons), cathode rays, positive rays, Lenard rays, magnetic effects, spark and arc discharge. A mathematical treatment with descriptions of theories, apparatus, experiments and results. Many source refs. in the text; titles not given.

731 Marchant, Edgar W., "The Delineation of Alternating Current Wave Forms." Dictionary of Applied Physics, Vol. 2, pp. 33-43. 19 illus. Survey of electromechanical instruments with mention of the cathode-ray oscillograph. 18 footnotes. See also pp. 30-32. See 51.

732 Johnson, John B., "A Low Voltage Cathode Ray Oscillograph." B. S. T. J., 1:142-151, Nov. 1922. 7 illus. Description of a hot-cathode, gas-focused tube operating at 300-400 volts, with electrostatic deflection plates.

733 MacGregor-Morris, John T., and R. Mines, "Measurements in Electrical Engineering by Means of Cathode-Rays." J. IEE., 63:1056-1107, Nov. 1925. 54 illus., 8 tables. App., pp. 1096-1105, "The jet as a measuring device," by R. Mines. Bibliography, pp. 1105-1107, 125 books and articles referred to in the text. Titles not given. This "survey of the present state of knowledge" is a comprehensive review from the initial investigations of Julius Plücker in 1858. Pt. 1, Sect. 3, pp. 1066-1074, is largely historical up to 1900. Pt. 2, pp. 1074-1095, on the cold-cathode instrument, hot-cathode instrument, methods of focusing, indicating and recording, and time scales. A basic historical paper.

734 Johnson, John B., "The Cathode Ray Oscillograph." J. F. I., 212:687-717, Dec. 1931; B. S. T. J., 11:1-27, Jan. 1932. 36 photos and diags., 29 footnote refs. from 1894. Theory, practice, operation and applications of the low-voltage cathode-ray tube. Surveys history and development, with details of tube construction, including high-voltage oscillographs.

735 Watson-Watt, Robert A., J. F. Herd, and L. H. Bainbridge-Bell, Applications of the Cathode-Ray Oscillograph in Radio Research. London: H. M. S. O., 1933. 290 pp., illus. 50 refs. from 1897. Mainly on equipment, circuits, and techniques developed to study atmospherics and the ionosphere, including a cathode-ray direction finder. Historical background, pp. 3-7, on early work at the British Meteorological Office from 1915, later in the Department of Scientific and Industrial Research. Other historical matter throughout that is pertinent to the early history of radar as well as cathode-ray tubes and oscilloscopes, waveform measurements and recordings.

736 Parr, Geoffrey, "The History of the Cathode-Ray Tube." Television and Short-Wave World, 10:85-87, Feb. 1937. 7 illus. "An account of how it was developed." A short survey, not documented.

737 MacGregor-Morris, John T., "The History and Development of the Cathode Ray Tube." J. Television Soc., 2(Ser. 2):257-262, June 1937. 6 illus. Brief mention of highlights, no refs.

738 Parr, Geoffrey, The Low Voltage Cathode-Ray Tube and Its Applications. London: Chapman and Hall, 1937. 177 pp., illus. Comprehensive bibliography, pp. 157-174, 823 entries from 1905. Emphasizes applications and measurements. Also, Parr, G., and O. H. Davie, The Cathode-Ray Tube and Its Applications. London: Chapman and Hall, 1952, 3rd ed. 1959, 433 pp.

739 Rettenmeyer, Francis X., "Radio-Electronic Bibliography." Radio, pp. 40, 42, 44, Dec. 1943. 13, Cathode-Ray Oscillographs, 156 items from 1918. See 93.

740 Puckle, Owen S., Time Bases (Scanning Generators): Their Design and Development. With Notes on the Cathode Ray Tube. London: Chapman and Hall; New York: John Wiley, 1944. 204 pp., 124 illus. App. 1, pp. 139-

160, topics concerning the cathode ray tube. 6 other Apps., pp. 161-197, treat other topics related to time bases. 83 footnotes, mainly refs. to papers and patents. Bibliography, p. 198, 4 books and 11 papers. 2nd ed., 1951. 387 pp., over 300 refs. from 1893. Bibliography, pp. 377, 378. A standard text with historical background.

741 Ardenne, Manfred von, "Evolution of the Cathode-Ray Tube. A Survey of Developments over Three Decades." Wireless World, 66:28-32, Jan. 1960. 7 illus., 21 refs. in 18 entries. A general survey of cathode-ray tubes in early television, and in oscillographs for special purposes, with emphasis on the author's pioneer contributions from 1928.

742 Schlesinger, Kurt, and Edward G. Ramberg, "Beam-Deflection and Photo Devices." Proc. IRE., 50:991-1005, May 1962. 19 illus. A thorough survey of beam tubes for television, oscilloscopes, image storage and display; phototubes, image tubes and light amplifiers up to 1960. 6 footnotes, 163 refs. from 1839.

743 Bunshah, Rointan F., "The History of Electron Beam Technology." In, Robert Bakish (ed.), Introduction to Electron Beam Technology. New York: John Wiley, 1962. Chap. 1, pp. 1-20. 7 illus. Brief survey of highlights from Hauksbee's experiments of 1705 to electron optics, and development of electron beam heating equipment for industrial applications. 69 refs. from 1705.

E. Photoelectricity

Discoveries of the electrical effect of light on selenium and metal surfaces promoted new theories concerning the nature of light, energy and matter. Further work after X rays and the electron were discovered led to revolutionary concepts in theoretical physics during the early decades of the twentieth century. General history, particularly from the scientific viewpoint, is recounted in a number of books listed in Sect. 3A, especially 197, 198, 200, 205, 212. Applications for light-sensitive devices, although attempted with selenium cells as early as the late 1870s, became practical after the introduction of phototubes and improved photocells around 1930. Related communication devices and systems up to the 1920s are surveyed in 1193, pp. 125-155. The expansion of interest in photoelectric devices during the early part of the 1930s is well shown by the changes in textbooks of the period. See also Sects. 7G, 8C, and Chap. 16 for developments in camera tubes, photomultipliers and related television and facsimile applications.

744 Smith, Willoughby, "The Action of Light on Selenium." J. Soc. Telegraph Engineers, 2:31-33, 1873; 5:183, 1876; Am. J. Sci., 5:301, 1873. See also, 1658, 1774.

745 Hertz, Heinrich R., "On an Effect of Ultra-Violet Light upon the Electric Discharge." Electric Waves, Chap. 4, pp. 63-79. 4 diags. Trans. from the German, Ann. der Phys., 31:983-1000, 1887. See 872.

746 Hallwachs, Wilhelm, "Electric Discharge by Light." Source Book in Physics, pp. 578, 579. Extracts from the German, originally published in Ann. der Phys., 33: 301-312, 1888. See 192.

747 Elster, Julius, and Hans Geitel (Oliver J. Lodge, trans.), "On the Diselectrification of Metals and Other Bodies by Light." Signalling Through Space Without Wires, App. 4, pp. 115-126. 8 diags. 12 abstracts from Ann. der Phys., Vols. 38 to 52, 1889-1894. See 874.

748 Einstein, Albert, "The Photoelectric Effect." Great Experiments in Physics. Chap. 17, pp. 232-237, including introduction. 5 refs. Brief description adapted from the original, Ann. der Phys., 17:132-148, 1905. See 212.

749 Allen, Herbert S., Photo-Electricity: the Liberation of Electrons by Light. With Chapters on Fluorescence and Phosphorescence, and Photo-Chemical Actions and Photography. London: Longmans, Green, 1913. 221 pp., 35 illus. A full survey with extensive historical background. Thoroughly documented with over 400 footnote refs. 2nd ed., 1925. 320 pp.

750 Hughes, Arthur L., Photo-Electricity. Cambridge: University Press, 1914. 144 pp., diags., tables.

751 Allen, Herbert S., "Photoelectricity." Dictionary of Applied Physics, Vol. 2, pp. 593-598. 62 footnote refs. from 1887, titles not given. List of refs., 10 items. See 51.

752 d'Albe, E. E. Fournier, The Moon-Element. An Introduction to the Wonders of Selenium. London: T. Fisher Unwin, 1924. 166 pp., 32 illus., including 6 plates. Selenium, selenium cells, relay applications, conversion of light into sound, the talking film. There is a short discussion of picture transmission and television, pp. 75-83. The inventor of the Optophone gives a history of its development, pp. 94-146, with details of demonstrations and efforts to promote it as a practical reading instrument for the blind.

753 Doty, Marion F. (comp.), Selenium 1817-1925: A List of References in the New York Public Library. New York Public Library, Bull. 30:440-448, 525-555, 599-629, 728-737, 793-824, June to Oct. 1926. An extensive collection of 1665 items chronologically arranged; bibliography and general works, constants, properties, chemistry, cells, industrial applications. Also published separately, 1927, 114 pp.

754 Campbell, Norman R., and Dorothy Ritchie, Photoelectric Cells: Their Properties, Use, and Applications. London: Pitman, 1929. 209 pp., 41 illus., footnote refs. 2nd ed., 1930, 217 pp. 3rd ed., 1934, 233 pp. An early practical text.

755 Barnard, George P., The Selenium Cell: Its Properties and Applications. London: Constable; New York: Richard R. Smith, 1930. 331 pp., 258 photos, diags., port. of Willoughby Smith. Chap. 1, pp. 3-17, Early history of selenium. Chap. 2, pp. 18-39, Construction of selenium cells. Chaps. 3, 4, pp. 40-157, cover the properties of selenium. Chap. 5, pp. 158-185, Theories of the action of light on selenium. Chaps. 6, 7, pp. 189-239, cover applications. Chaps. 8, 9, pp. 240-291, on the uses of selenium for light-telephony, talking films, facsimile and television. Chap. bibliographies contain over 800 entries, many multiple items. Apps., about 200 additional refs. A basic work.

756 [B. T. L.]. Bibliography of Articles on Photo-electricity, 1896-June 1930. New York: Bell Telephone Laboratories, 1930. 145 pp.

757 Zworykin, Vladimir K., and E. D. Wilson, Photocells and Their Application. New York: John Wiley, 1930. 209 pp., 98 illus. 114 chap. refs., 36 other refs., bibliography, 23 items. 2nd ed., 1932. 331 pp., 180 illus. Revised, 1934. 348 pp., 180 illus. 195 chap. refs., 55 cited

refs., bibliography, 29 items. Chap. 1, pp. 1-14, Historical introduction, other historical matter in the text. Electronic television (von Ardenne, Farnsworth, Zworykin) is covered in pp. 274-293.

758 Nix, Foster C., "Photo-conductivity." Rev. Mod. Phys., 4:723-766, Oct. 1932. 33 diags. and graphs. Photoconductivity in insulators and semiconductors. Bibliography, pp. 759-766, 189 entries, mainly of the 1920s.

759 Hughes, Arthur L., and Lee A. DuBridge, Photoelectric Phenomena. New York: McGraw-Hill, 1932. 531 pp., 481 illus., 41 tables. Short book list, p. 6, 14 items. Extensively documented with about 1350 footnotes, other footnotes with some tables. Titles of articles not given. A basic work.

760 Walker, R. C., and T. M. C. Lance, Photoelectric Cell Applications. London: Pitman, 1933. 2nd ed., 1935. 3rd ed., 1938. 336 pp., 200 photos, graphs, diags. "A practical book describing the uses of photoelectric cells in television, talking pictures, electrical alarms, counting devices, etc." Also advertising, phototelegraphy, scientific instruments. 78 chap. refs.

761 Koller, L. R., "Photoelectricity." The Physics of Electron Tubes, Chap. 11, pp. 141-165. 13 graphs and diags., 1 table. 17 refs. "Photoconductivity," Chap. 12, pp. 166-176. 7 graphs and diags., 2 tables, 9 refs. "The Photovoltaic Effect," Chap. 13, pp. 177-181. 3 graphs and diags. 7 refs. See 996.

762 Henney, Keith, "Light-Sensitive Tubes." Electron Tubes in Industry, Chap. 5, pp. 299-330. 30 photos, diags., and graphs. 11 footnote refs., bibliography, 17 items. Chap. 6, "Applications of Light-Sensitive Tubes," pp. 331-460. 84 photos, diags. and graphs. 64 footnote refs., bibliography, 18 items. See 1017.

763 [B. T. L.]. Bibliography of Photoelectricity. New York: Bell Telephone Laboratories, 1940. 218 pp. 1698 entries chronologically arranged and grouped alphabetically by author. Technical and theoretical articles from 1913.

764 Glover, Alan M., "A Review of the Development of Sensitive Phototubes." Proc. IRE., 29:413-423, Aug.

1941. 12 illus. 73 footnote refs. from 1839, bibliography, 10 items. Survey of developments from the late 1880s.

765 Stranathan, James D., "Photons--The Photoelectric Effect--Radiation and Absorption." The "Particles" of Modern Physics, Chap. 6, pp. 212-259. 14 photos and diags., 5 tables. 83 footnote refs. from 1887. See 688.

766 [B. T. L.]. Photoelectric Cells: Applications, 1913-1942. New York: Bell Telephone Laboratories, 1942. 229 pp. 1939 entries in chronological order. Bibliography of articles.

767 Zworykin, Vladimir K., and E. G. Ramberg, Photoelectricity and Its Application. New York: John Wiley, 1949. 494 pp., illus. Vacuum and gas-filled phototubes, photo-multipliers, image tubes, photoconductive and photovoltaic cells. Extensive coverage includes camera tubes. Chap. bibliographies, about 450 items from 1834. A standard text.

768 Summer, W., Photosensistors: A Treatise on Photoelectric Devices and Their Application to Industry. London: Chapman and Hall, 1957; New York: Macmillan, 1958. 675 pp., illus. Photocells, measurements, remote control. Well documented with 2650 refs. to articles and books from 1823. Lists 687 patents.

F. Electron Optics and the Electron Microscope

The early history of the electron microscope is embodied in the development of high-voltage oscillographs, used to investigate the nature of fast electrical transients and thunderstorm effects on power transmission lines, and the concurrent work on electron optics during the 1920s. For related literature, see 686, 687, electron waves; 733, cathode-ray oscillographs; and Sect. 16C for electron optics applied in the design of television camera and receiver tubes, particularly 1739, 1748, 1752.

769 Wood, A. B., "Recent Developments in Cathode-Ray Oscillographs." J. IEE., 71:41-56 (disc. 70-82), June 1932. 16 photos, diags., and graphs. High-voltage oscillographs, films, screens, internal and external photography, the Lenard window, time recording, time axis and

timing circuits, potential dividers. 33 refs. from 1918.

770 Zworykin, Vladimir K., "Electron Optics." J. F. I., 215:535-555, May 1933. 16 photos and diags. 6 footnote refs.

771 Miller, J. L., and J. E. L. Robinson, "The High-Speed Cathode-Ray Oscillograph." Rep. Prog. Phys., 2: 259-283, 1935. 12 photos and diags. On general construction, discharge tubes, photography, beam release and withdrawing systems, time delineation and bias controls, tripping and transient synchronization, voltage dividers, measurement of surge current. Refs., pp. 296-298, 77 items from 1919.

772 Starks, H. J. H., "The Electron Microscope." Rep. Prog. Phys., 2:283-291, 1935. Condensed review with a brief summary of work on electron optics. Refs., pp. 298, 299, 55 items from 1922. Titles not given.

773 Klemperer, Otto (comp.), Electron Optics. Cambridge: University Press, 1939. 107 pp., illus. By the research staff of Electric and Musical Industries, Ltd. Literature refs., pp. 102-104.

774 Myers, Leonard M., Electron Optics: Theoretical and Practical. London: Chapman and Hall, 1939. 618 pp., plates, diags., graphs. Comprehensive, with much historical material, including applications. Extensive bibliography, pp. 585-608, 800 entries from 1817.

775 Burton, Eli F., and Walter H. Kohl, The Electron Microscope: An Introduction to its Fundamental Principles and Applications. New York: Reinhold, 1942. 233 pp., illus., diags. General bibliography, pp. 227-228. 2nd ed., 1946. 325 pp., 125 diags., 96 photos, including some views of early apparatus. General introduction for the layman with historical matter in most chapters. Bibliography, pp. 299-318, 597 items from 1926, plus 24 items (reprinted from J. App. Phys.). First book in English on the subject.

776 Rüdenberg, Reinhold, "The Early History of the Electron Microscope." J. App. Phys., 14:434-436, Aug. 1943. Personal account of the original conception, with excerpts from two U. S. patents. 5 diags. 3 refs.

777 Marton, Claire, and Samuel Sass, "A Bibliography of

Electron Microscopy." J. App. Phys., 14:522-531, Oct. 1943; 15:575-579, Aug. 1944; 16:373-377, July 1945. 667 books and articles subdivided by topic in 8 categories. Supplement by Mary E. Rathbun and others, 17:759-762, Oct. 1946, 138 entries including 70 patents.

778 Gabor, Dennis, The Electron Microscope: Its Development, Present Performance and Future Possibilities. London: Hulton Press, 1945; revised American ed., Brooklyn, N.Y.: Chemical Publishing Co., 1948. 164 pp., 54 photos, graphs, diags. This monograph ranges from an introduction to theory and instruments to a more advanced treatment intended for engineers and physicists. Includes much historical material. Bibliography, pp. 151-157, 3 books and 85 refs., some multiple entries, from 1910.

779 Calbick, Chester J., "Historical Background of Electron Optics." J. App. Phys., 15:685-690, Oct. 1944. 13 photos and diags. Brief survey from 1897, with mention of work by C. J. Davisson and the author from 1929. Bibliography, 15 items from 1897.

780 Hawley, Gessner C., Seeing the Invisible. The Story of the Electron Microscope. New York: Alfred A. Knopf, 1945. 205 pp., 72 illus. Chap. 5, pp. 83-102, History of the electron microscope. No refs.

781 Zworykin, Vladimir K., and others, Electron Optics and the Electron Microscope. New York: John Wiley, 1945. 766 pp., 578 illus. The first half is practical, the remainder theoretical. A standard textbook with 394 items in chap. refs.

782 Cosslett, Vernon E., Introduction to Electron Optics: The Production, Propagation and Focusing of Electron Beams. Oxford: Clarendon Press, 1946. 272 pp., plates, diags. 169 chap. refs., bibliography, p. 263. Full coverage includes production of electron beams, the cathode-ray tube and its derivatives, electron diffraction, electron microscope, and other applications. 2nd ed., 1950, 293 pp., 8 plates, 159 diags. Reprint of 1st ed., with corrections and additional matter as chapter notes, pp. 263-279. Bibliography, p. 282, 12 items.

783 Cosslett, Vernon E., "Electron Microscopy and Electron Diffraction." Electronics and Their Application in Industry and Research, Chap. 14, pp. 535-641. 84 photos,

diags. and graphs. 106 refs. from 1926, booklist, 7 items, pp. 638-641. A technical survey of design, construction, limitations, and applications, with details of specific instruments. Historical refs. See 799.

784 Cosslett, Vernon E. (ed.), Bibliography of Electron Microscopy. London: Edward Arnold, 1950. 350 pp. About 2500 entries, plus cross-refs., alphabetically by author. Description of contents, with sources of abstracts. For the Institute of Physics.

785 Marton, Claire, and others, Bibliography of Electron Microscopy. Washington: National Bureau of Standards, 1950. Circular 502, 87 pp. 1923 entries from 1926.

786 Cosslett, Vernon E., "Electron Microscope--Past, Present and future." J. Roy. Soc. Arts, 110:668-688, Aug. 1962. Illus.

787 Mulvey, T., "Origins and Historical Development of the Electron Microscope." Brit. J. App. Phys., 13: 197-207, May 1962. 20 photos and diags. Early speculations, cathode-ray tubes, electron optics, focusing devices, practical problems, commercial instruments and applications. 70 refs. from 1858, titles not given.

788 Freundlich, Martin M., "Origin of the Electron Microscope." Science, 142:185-188, Oct. 11, 1963. 2 illus. On development work by Max Knoll and Ernst Ruska, and R. Rüdenberg's part in the development. 35 refs., notes and patents from 1878.

789 Mulvey, T., "The History of the Electron Microscope." Proc. Roy. Microscopic Soc., 2(Pt. 1):201-227, 1967.

790 Marton, Ladislaus, Early History of the Electron Microscope. San Francisco: San Francisco Press, 1968. 56 pp., 11 illus. Preface by Dennis Gabor. A personal account of original research from 1932 up to the mid-1940s, with details of instruments, techniques and the contributions of other scientists. Bibliographical note, p. 50, 9 entries. Refs., pp. 51-55, 102 entries (some multiple) from 1733. 39 entries in English.

G. Electronics

Most of the entries in this section are reviews of progress and applications and therefore cover a rather wide range of subjects. The M. I. T. Radiation Laboratory Series (798), although centered on radar and microwaves, is included herein because the set is practically an encyclopedic record of electronics progress through the years of World War II. See also Sects. 3E, 3G for related titles in telecommunications and the electronics industry; 305, 309, 340, 341, 343.

791 Süsskind, Charles, "The Origin of the Term 'Electronics.'" IEEE Spectrum, 3:72, 77-79, May 1966. 2 illus. Brief review of literature, theories, terminology and usages. 37 refs. from 1832.

792 Yates, Raymond F., These Amazing Electrons. New York: Macmillan, 1937. 326 pp., 46 plates, 63 diags. A comprehensive survey of electronic applications with a popular and romantic treatment. A little history, no refs.

793 Henney, Keith, "A Decade of Electronics 1930-1940." Electronics, 13:17-24, April 1940. This compact review is almost a catalog of progress centered on the magazine's history. The first five years are covered in some detail, with statistics of tube and receiver production and miscellaneous references to technical matters. There are numerous references to the month of issue with respect to articles and reports.

794 Windred, George, "Review of Progress in Electronics." Electronic Engineering, Vol. 14, April-Dec. 1941. 1, Introduction and general bibliography, 151, 152, 154. 2, Photo-electrolytic effects, 209-211. 3, Photoconductivity, 249-251, 256, 270. 4, Photo-voltaic effects, 298-300, 321. 5, Photo-electricity, 345-347. 6, Thermionic emission, 391, 392. 7. Electrical conduction in gases, 442, 443, 465. 8, Grid control of gaseous conduction, 487-489. 9, Measurement of the electron, 538, 539, 561. Experimental techniques and practical applications. 31 photos, graphs and diags. Bibliography (with each issue) totals 116 items.

795 Mills, John, "Forty Years of Electronics." Tech.

Rev., 46:486-488, June 1944. Illus.

796 Friedlander, Gordon D., "World War II: Electronics and the U.S. Navy." IEEE Spectrum, 4:56-70, 46-56, Nov., Dec. 1967. On radar, sonar, loran, and infrared techniques. 26 illus., 6 refs., 4 entries in the bibliography.

797 Stokley, James, Electrons in Action. New York: McGraw-Hill, 1946. 320 pp., 36 plates, 55 diags. On contemporary electronic devices and applications with some historical material. App., pp. 293-309, on radar, with 4 plates. Occasional footnotes, numerous names, dates, patent numbers and extracts in the text. A popular treatment.

798 [M. I. T.]. Radiation Laboratory Series. New York: McGraw-Hill, 1947-1953. 28 vols., including index vol. A comprehensive record of technical developments during World War II. Although there is much overlapping of main subjects, the following is an approximate grouping: microwaves and systems, 12; radar and navigation, 4; instruments and measurements, 3; servos, 2; miscellaneous, 6. About 15,850 pp., 8000 illus. Refs. and bibliographies.

799 Lovell, Bernard (ed.), Electronics and Their Application in Industry and Research. London: Pilot Press, 1947. 660 pp., 404 illus., 13 tables. On recent advances in particular fields, but some chapters contain historical material. Introduction, pp. 1-11; Photocells, pp. 53-95; Television, pp. 135-211; Thermionic valves for very high frequencies, pp. 213-237; Radar, pp. 239-319; High frequency heating, pp. 349-381. Other chapters on electron physics, servomechanisms, medical electronics, the Betatron, electron microscopy, and various specialized topics. 520 chap. refs., 33 items in 3 chap. bibliographies.

800 Berkner, Lloyd V., "Electronics Comes of Age." Elect. Eng., 67:32-37, Jan. 1948. Photo. An overview; on the emergence of new industries, research and development, and industrial systems.

801 White, William C., "Electronics--Past, Present and Future." J. F. I., 248:367-379, Nov. 1949. 5 photos. Chronology showing basic researches, products and dates from 1725. Grouped as high-vacuum tubes, electron or ion

beams in high vacuum, phototubes, cold-cathode discharge tubes, mercury pool cathode tubes, and hot-cathode, glow-discharge tubes.

802 Andres, Paul G., Survey of Modern Electronics. New York: John Wiley; London: Chapman and Hall, 1950. 522 pp., 355 illus. A comprehensive view of contemporary electronic devices and applications at the threshold of the transistor age. Numerous minor historical items in the text. Bibliography, pp. 501, 502, 24 general refs., list of periodicals and film sources. Extensive ref. list with each chap.

803 Pierce, John R., "Electronics." Sci. Am., 183:30-39, Oct. 1950. 15 illus. General survey of electron physics and electron tubes from 1860. No refs.

804 Nottingham, Wayne B., Bibliography on Physical Electronics. Cambridge, Mass.: Addison-Wesley, 1954. 428 pp. Wide coverage of books and articles in classified groups; gaseous, surface emission, solid state, phosphors and luminescence, photoelectric, techniques, miscellaneous. Prepared by the staff of the Research Laboratory of Electronics, Massachusetts Institute of Technology. Covers 1930-1950.

805 Zworykin, Vladimir K., "Some Prospects in the Field of Electronics." J. F. I., 251:69-80, Jan. 1951. 8 photos, 7 refs. Also, Smithsonian Institution, Annual Report for 1951, pp. 235-243 (Washington, 1952). Brief survey of the past and a discussion of the electron microscope, electronic computers, television, semiconductors, and future trends.

806 Lessing, Lawrence P., "The Electronics Era." Fortune, 44:78-83, 132, 134-136, 138, July 1951. 12 photos and diags. List of 20 companies. On the TV boom, military electronics, transistors, computers.

807 Hobson, Jesse E., "Electronics, Development, and Industrial Growth." Elect. Eng., 71:986-991, Nov. 1952.

808 Zeluff, Vin, and William G. Arnold, "25 Years of Electronics." Electronics, 28:124-129, April 1955. 6 charts showing growth of the industry, 2 short lists of highlights of early radar and industrial applications.

809 Fink, Donald G., "Electronic Developments in the

United States." J. IEE., 4(NS):533-540, Oct. 1958. 6 illus. Brief survey of developments since 1946. No refs.

810 Quarles, Donald A., "Military Electronic Developments and Their Applications." Elect. Eng., 78:435-448, May 1959. 19 photos, diags., and graphs. This view over 25 years includes radar, data processing, aircraft electronics, computers, guided missiles, communications, miniaturization, and engineering research and development. No refs.

811 "Signal Corps Centennial Issue." Trans. IRE., MIL-4, Oct. 1960. 35 papers (pp. 396-607) on high-frequency communications, radar, and developments in military electronics, including some historical surveys.

812 Overhage, Carl F.J. (ed.), The Age of Electronics. New York: McGraw-Hill, 1962. 218 pp., 138 illus. A collection of essays on selected topics, some with historical background. Chap. 1, Introduction; Chap. 2, Maxwell, Hertz, and Lorentz; Chap. 4, Radar; Chap. 8, Masers. 75 chap. refs.

813 Spielman, Harold S., Electronics Source Book for Teachers. New York: Hayden, 1965. 3 vols., illus. Numerous refs. for each chap., primarily 1955-1964 sources. Useful coverage for the period, with a sprinkling of earlier history.

814 Kompfner, Rudolf, "Electron Devices in Science and Technology." IEEE Spectrum, 4:47-52, Sept. 1967. 2 illus. A brief survey of notable developments in electron devices from 1890 to the 1960s. Also, Dec., pp. 139, 140, 2 letters with 16 and 7 refs.

815 Handel, Samuel, The Electronic Revolution. Harmondsworth, Middlesex; Baltimore: Penguin Books, 1967. 252 pp., 59 illus. List of books for further reading, pp. 246-248, 29 items. Popular review of the growth of electronic technology with a general historical background. No refs.

816 Süsskind, Charles, "The Early History of Electronics." IEEE Spectrum. 1, Electromagnetics before Hertz, 5:90-98, Aug. 1968. 2, The experiments of Hertz, 5:57-60, Dec. 1968. 3, Prehistory of radiotelegraphy, 6:69-74, April 1969. 4, First radiotelegraphy experiments, 6:66-70, Aug. 1969. 5, Commercial beginnings, 7:78-83, April 1970. 6,

Discovery of the electron, 7:76-79, Sept. 1970. Ports., extensive refs. In progress.

817 Shore, Bruce H., The New Electronics. New York: McGraw-Hill, 1970. 253 pp., 91 photos. Selected bibliography, pp. 231-237, 103 items, mainly books and articles published during the 1960s. A popular exposition of advanced topics in modern physics and solid-state electronics; magnetic, crystal and thin-film materials, computer memories, superconductivity, millimeter waves, electronic photography, lasers and holography. Semitechnical explanations are given and there is some brief historical material. No refs.

8. ELECTROACOUSTICS

The introduction of the telephone in the mid-1870s marks the beginning of practical electroacoustics. See Sect. 15C for telephone history, also Chap. 3, particularly 238, 247, 261, 287. Earlier history is covered in 818, which includes copious references to leading papers and patents concerning telephones, microphones, loudspeakers, and systems of particular interest to audio specialists.

A. General

818 Hunt, Frederick V., Electroacoustics: The Analysis of Transduction, and Its Historical Background. Cambridge, Mass.: Harvard University Press; New York: John Wiley, 1954. 260 pp., 52 illus. Index of names, pp. 247-253, includes dates. Chap. 1, pp. 1-91, Historical context, is a thorough survey with 251 footnote refs. from 1558 up to about 1940. Many multiple entries and patent numbers. A key to the location of rare items is included. English translations not given Footnote refs. total 303.

819 Boyle, Robert W., "Ultrasonics." Science Progress, 23:75-105, July 1928. Report on research and development in Britain during World War I and later.

820 Blake, George G., "Various Types of Microphones." History of Radio Telegraphy and Telephony, Chap. 13, pp. 194-209. 16 illus. See 1193.

821 Frederick, Halsey A., "The Development of the Microphone." J. Acoustical Soc. America, 3:Supp., July 1931. 25 pp., 24 illus. Also, Bell Tel. Qtly., 10:164-188, July 1931.

822 Tournier, Marcel C., "History and Application of Piezoelectricity." Elect. Comm., 15:312-327, April 1937. A review of research, particularly of the work of the Curie brothers, W. Voigt, P. Langevin, and ultrasonic applications.

823 Bergmann, Ludwig (H. S. Hatfield, trans.), Ultrasonics and Their Scientific and Technical Application. London: G. Bell, 1938. 264 pp., illus. A standard work. 575 refs. from 1847.

824 Rettenmeyer, Francis X., "Radio-Electronic Bibliography." Radio, pp. 26, 29, 30, 32, 34, 35, 39, 40, Oct. 1942. 6, Filters, Sound, Loudspeakers, 326 items from 1915. See 93.

825 Jupe, J. H., "Ultrasonics: a Brief Survey." Electronic Engineering, 21:422-424, Nov. 1949. Includes history as well as applications. 115 refs.

826 Vigoureux, Paul, Ultrasonics. London: Chapman and Hall, 1950. 163 pp., 73 illus., 25 tables. Bibliography, pp. 145-157, 328 entries from 1894.

827 Olson, Harry F., Acoustical Engineering. Princeton, N. J.: D. Van Nostrand, 1957. 718 pp., 569 illus., 11 tables. Covers all aspects of theory and practice. Includes a survey of the means for the communication of information, Chap. 14, pp. 657-668. Chap. 15, pp. 669-691, Underwater sound. Chap. 16, pp. 692-703, Ultrasonics. 848 footnote refs. Subject index only. A basic work.

828 Bauer, Benjamin B., "A Century of Microphones." Proc. IRE., 50:719-729, May 1962. 7 illus. 73 footnote refs. from 1820.

829 "Audio Pioneers." Audio, 46:32-60, 72, 73, May 1962. 35 biographical essays with ports.

830 McProud, C. G., "20 Years of Audio." Audio, 51:25-28, 32, 34, 36, May 1967. 32 photos. A review by the publisher.

831 Sivowitch, Elliot N., "Musical Broadcasting in the 19th Century." Audio, 51:19-23, June 1967. 15 illus. On experiments with telephone installations for public concerts from the mid-1870s.

B. Recording and Reproduction

832 Rice, Chester W., and Edward W. Kellogg, "Notes on the Development of a New Type of Hornless Loud

Speaker." Trans. AIEE., 44:461-475, April 1925. 34 photos, graphs and diags. Discussion, pp. 475-480. 18 footnote refs. A basic paper.

833 Kellogg, Edward W., "Electrical Reproduction From Phonograph Records." J. AIEE., 46:903-911 (discussion, 911, 912), Feb. 1927. 9 photos, charts, diags. Bibliography, 15 items, amplifiers and loud speakers. 10 footnote refs.

834 Wilson, Percy, and G. W. Webb, Modern Gramophones and Electrical Reproducers. London: Cassell, 1929. 271 pp., 120 photos and diags. An introduction and historical survey. Bibliography, pp. 257-265, includes many British patents, 160 items from 1876.

835 Begun, Semi J., Magnetic Recording. New York: Murray Hill Books, 1949. 242 pp., illus. Chap. 1, pp. 1-15, History of magnetic recording. Bibliography, pp. 12-15, lists 43 articles and 11 patents. 188 refs. from 1888. A basic book.

836 Read, Oliver, The Recording and Reproduction of Sound. Indianapolis: Howard W. Sams, 1949. 364 pp., 256 illus. Chap. 2, pp. 10-17, History of acoustical recording, 11 illus. Bibliography of magnetic recording, pp. 315-318, 66 entries from 1927.

837 Wilson, Carmen F., Magnetic Recording, 1900-1949. Chicago: John Crerar Library, 1950. 61 pp. 339 annotated entries, including patents.

838 Wilson, Carmen F., "Magnetic Recording, 1888-1952." Trans. IRE., AU-4:53-81, May-June, 1956. List of 38 important patents from 1899. 650 items arranged by year from 1888, also 10 miscellaneous items. Most entries annotated. A complete bibliography.

839 Snyder, R. H., "History and Development of Stereophonic Sound Recording." J. Audio Engineering Society, 1:176-179, April 1953.

840 [RCA]. The 50-Year Story of RCA Victor Records. New York: Radio Corporation of America, 1953. 77 pp., highly illustrated, mainly with photos of artists and studio scenes. Contains a little history; Triumphs of yesterday, pp. 14-37.

841 Jorysz, Alfred, "Bibliography of Magnetic Recording, 1900-1953." J. Audio Engineering Society, 2:183-199, July 1954. Also Tele-Tech, 13:54-56, 111, 112, July 1954. 353 items.

842 Gelatt, Roland, The Fabulous Phonograph. From Edison to Stereo. New York: J. B. Lippincott, 1955. Rev. ed., Appleton-Century, 1965. 336 pp., 16 plates, 8 illus. Chronology, pp. 321-326, 52 entries from 1877. Largely about companies, promoters, legal and business battles, growth of the industry, artists and their associates in the music field.

843 Spencer, Kenneth J., High Fidelity: a Bibliography of Sound Reproduction. London: Iota Services, 1958. 325 pp. Wide coverage, 2700 items, many annotated.

844 Tall, Joel, Techniques of Magnetic Recording. New York: Macmillan, 1958. 472 pp. Brief history from 1900, with descriptions of applications. Bibliography, about 250 refs.

845 Read, Oliver, and Walter L. Welch, From Tin Foil to Stereo: Evolution of the Phonograph. Indianapolis: Howard W. Sams; New York: Bobbs-Merrill, 1959. 524 pp., 185 photos, graphs, diags., including port. (Edison), and 16 plates. On men, inventions, patents, companies and products, with many extracts and critical comments. Chap. 16, pp. 219-236, superficial mention of selected events in early radio up to 1912. Chaps. 19, 20, pp. 275-288; 289-299, on motion pictures and sound recording. Corporate genealogy chart, pp. 486-489. Bibliography, pp. 495-502, 240 items in 3 categories. 120 footnotes include some refs.

846 Sunier, John, The Story of Stereo: 1881-. New York: Gernsback, 1960. 160 pp., 81 photos and diags. Chap. 2, pp. 27-50, Early developments, is a survey from 1881. Historical background in other chapters on film, tape, disc, and broadcasting. Chapter bibliographies contain 407 refs., mostly to pages in articles.

847 Bachman, William S., Benjamin B. Bauer, and Peter C. Goldmark, "Disk Recording and Reproduction." Proc. IRE., 50:738-744, May 1962. Table. History and highlights of development from 1878. 58 footnote refs. from 1887.

848 Hilliard, John K., "The History of Stereophonic Sound Reproduction." Proc. IRE., 50:776-780, May 1962. Primarily a condensed, narrative bibliography. 121 refs. in 8 groups, mostly from about 1940.

849 Tinkham, Russell J., "Anecdotal History of Stereophonic Recording." Audio, 46:25-31, 91, 93, May 1962. 8 illus. Personal reminiscences, with quotes and extracts. 9 refs. from 1926.

C. Sound Motion Pictures

850 de Forest, Lee, "The Phonofilm." Trans. SMPE., 16:61-75, May 1923. Description of early optically recorded sound film.

851 de Forest, Lee, "The Motion Picture Speaks." Popular Radio, 3:159-169, March 1923. 6 illus. On the author's Phonofilm process and the "photion" cell. Some historical background.

852 Ramsaye, Terry, "Early History of Sound Pictures." Trans. SMPE., 31:597-602, Sept. 1928.

853 Green, Fitzhugh, The Film Finds Its Tongue. New York; London: G. P. Putnam's Sons, 1929. 316 pp., 31 plates. On early talking pictures and Warner Brothers. No refs. No index.

854 Miehling, Rudolph, Sound Projection. New York: Mancall, 1929. 528 pp. 137 photos and diags. Chap. 1, pp. 1-8, Historical data. In addition to the basics of sound and electricity, there are chapters on general topics such as studio and recording techniques, light-sensitive cells, sound pick-up devices, vacuum valves, amplifiers, sound reproducers, power supplies, battery chargers, and auditorium acoustics. Twelve chapters treat operating and related procedures. Five chapters deal in detail with specific equipment and systems. A useful contemporary record.

855 Cowan, Lester (ed.), Recording Sound for Motion Pictures. New York: McGraw-Hill, 1931. 404 pp., 4 plates, 236 illus., many of early equipment. 24 chapters by individual contributors, mostly on contemporary practices, but with some historical matter. Chap. 1, pp. 1-12, The ancestry of sound pictures, by H. G. Knox. In 4 parts;

sound recording equipment, the film record, studio acoustics and technique, sound reproduction. No refs.

856 Crawford, Merritt, "Some Accomplishments of Eugene Augustin Lauste--Pioneer Sound-Film Inventor." J. SMPE., 16:105-109, Jan. 1931. Also 17:632-644, Oct. 1931.

857 Wente, E. C., "Contributions of Telephone Research to Sound Pictures." J. SMPE., 27:188-194, Aug. 1936.

858 Theisen, W. E., "Work of Lee de Forest." J. SMPE., 35:542-548, Dec. 1940.

859 de Forest, Lee, "Pioneering in Talking Pictures." J. SMPE., 36:41-49, Jan. 1941.

860 Theisen, W. E., "Pioneering in the Talking Picture." J. SMPE., 36:415-444, April 1941.

861 Lovette, Frank H., and Stanley Watkins, "Twenty Years of Talking Movies, an Anniversary." Bell Tel. Mag., 25:82-100, Summer, 1946.

862 Sponable, E. I., "Historical Development of Sound Films." J. SMPE., 48:275-303, 407-422, April, May 1947. A review from 1857 to 1926, with bibliography and list of patents.

863 Kellogg, Edward W., "History of Sound Motion Pictures." J. SMPTE., 64:291-302, 356-374, 422-437, June, July, Aug. 1955. 17 illus. A thorough survey, primarily of American developments from the 1920s. Bibliography and refs. with each part total 406 items from 1878. Numerous multiple entries, many patents included. A basic paper.

864 Batsel, Max C., and Glenn L. Dimmick, "Film Recording and Reproduction." Proc. IRE., 50:745-751, May 1962. 11 illus. Historical summary from 1926.

9. ELECTROMAGNETIC WAVES

Entries in this chapter cover early experiments, discoveries and theories concerning electromagnetic radiation and the evolution of antennas and propagation theory. Kindred subjects included deal with theories of the ether, resonance, interference, transmission lines, microwaves, waveguides, modulation, regulations, and radio astronomy. Other relevant entries will be found in Sect. 7A; 645, 650, 651, 653; and in Sects. 11B, 12A, 13B.

A. Observations and Experiments to 1900

865 Maxwell, James C., "A Dynamical Theory of the Electromagnetic Field." Proc. Roy. Soc., 23:531-536, 1864; Phil. Trans., 155:459-512, 1865; Phil. Mag., 29:152-157, 1865. Also in Scientific Papers, Vol. 1, pp. 526-571. Pt. 1, Introductory, pp. 459-466 (Phil. Trans.), is a non-mathematical explanation of the theory that links electromagnetic radiations ("and other radiations if any") with light. The rest of the paper develops the general equations of the electromagnetic field. One of the monumental papers of the 19th century. See also 192, pp. 528-538.

866 Jehl, Francis, "Etheric Force: The Famous Black Box." Menlo Park Reminiscences, Vol. 1, Chap. 12, pp. 80-88. 2 photos, 2 diags. Reprint of a page from Edison's notebook and of an article in The Operator, Jan. 1876. An account of experiments by Edison with unknown radiations. See 534, also 867-870, and 1223.

867 Dunlap, Orrin E., "Edison Glimpsed at Radio in 1875." Sci. Am., 135:424, 425, Dec. 1926. 3 illus.

868 Snyder, Monroe B., "Professor Elihu Thomson's Early Experimental Discovery of the Maxwell Electro-Magnetic Waves." G. E. Rev., 23:208, March 1920. On experiments of 1875 (repetition of an earlier experiment of 1871).

869 Goodman, Nathan G., "Earliest Wireless Experiments in America. Elihu Thomson Played with Idea over Half Century Ago." J. F. I., 224:311-314, Sept. 1937. Mainly a letter by Elihu Thomson about his experiments of 1875.

870 Süsskind, Charles, "Observations of Electromagnetic-Wave Radiation Before Hertz." Isis, 55:32-42, March 1964. Galvani, Henry, Edison, Thomson and Houston, S. P. Thompson, Dolbear, Hughes. 33 footnote refs. from 1791.

871 Fahie, John J., "Researches of Prof. D. E. Hughes, F. R. S., in Electric Waves and Their Application to Wireless Telegraphy, 1879-1886." A History of Wireless Telegraphy, 1838-1899, App. D, pp. 289-296. Mainly contents of a letter from Hughes to the author. See 1226.

872 Hertz, Heinrich R. (D. E. Jones, trans.), Electric Waves. London: Macmillan, 1893; New York: Dover, 1962. 279 pp., 40 illus., 127 footnotes. Supplementary notes, pp. 269-278. This classic work is subtitled: "Being researches on the propagation of electric action with finite velocity through space." See 192, pp. 549-561 for an extract of Chap. 11, On electric radiation.

873 Appleyard, Rollo, "Heinrich Rudolf Hertz." Pioneers of Electrical Communication, Chap. 5, pp. 109-140. 14 photos, ports., diags. Mainly on the famous experiments of 1886-1888. See 359.

874 Lodge, Oliver J., The Work of Hertz and Some of His Successors. London: "The Electrician" Printing and Publishing Co., 1894. 2nd ed., 1898, 58 pp. 3rd ed., 1900, retitled Signalling Through Space Without Wires. Being a Description of the Work of Hertz and His Successors. 133 pp., 73 illus., mainly diags. 43 footnotes, most refs., other refs. in the text. Lecture, pp. 1-42. Telegraphic applications, pp. 45-72, with patent refs. History of the coherer, pp. 73-87. Experiments of David Hughes, pp. 88-94. Branly's experiments, pp. 95-108. Photo-electric phenomena, pp. 115-128. Miscellaneous notes, pp. 43, 44, 109-114, 127-133. No index. Also, Electrician, 33:153-155; 186-190; 204, 205; 271, 272; 362, June 8-July 27, 1894, 25 illus. Nature, 50:133-139, June 1894. Based on a lecture by Lodge at the Royal Institution, June 1, 1894. Besides being a memorial to Heinrich Hertz, it also triggered interest in the initial experiments on wireless telegraphy. Various issues and paginations, with additions, deletions and changes in the

text and the appended material. 4th ed., 1908.

875 Thompson, Silvanus P., "The Invisible Spectrum." Light; Visible and Invisible, pp. 214-229. Also, "The Electromagnetic Theory of Light," pp. 230-237. On the work of Maxwell, Hertz, Lodge, Righi, Bose, and the author's early experiments in 1876. See 697.

876 Hyndman, Hugh H. F., Radiation; An Elementary Treatise on Electromagnetic Radiation and on Röntgen and Cathode Rays. London: Swan Sonnenschein; New York: Macmillan, 1898. 307 pp., 13 illus., several tables. The first 152 pages are largely devoted to the ether, ethereal vibrations, and early experiments and theories on electromagnetic waves. Vacuum tube phenomena, cathode rays and effects are treated in pp. 154-204. Pages 206-274 are on the Röntgen or X rays. 155 footnotes and numerous other refs. in the text, titles not given. This contemporary account of the exciting work of the 1890s omits mention of the practical work of Lodge, Marconi and others on early wireless communication experiments.

877 Whittaker, Edmund T., "Maxwell." A History of the Theories of Aether and Electricity, Vol. 1, Chap. 8, pp. 240-278. Also, "The Followers of Maxwell," Chap. 10, pp. 304-334. Copious footnote refs. See 659.

878 Hawks, Ellison, "D. E. Hughes and His Work." Pioneers of Wireless, Chap. 12, pp. 168-174. Also Chaps. 13, 14, pp. 175-188, 189-204, on Maxwell, Hertz, Righi, Branly, Lodge, and Popov. See 369.

879 Blake, George G., "Electromagnetic Waves." History of Radio Telegraphy and Telephony, Chap. 4, pp. 49-57. From Elihu Thomson's experiments to those by Chunder Bose, and Augusto Righi. Other parts on detectors and practical systems are also pertinent. See 1193.

880 Poincaré, Henri, "Maxwell's Theory and Hertzian Oscillations." Maxwell's Theory and Wireless Telegraphy, Pt. 1, pp. 1-107. Survey of work up to 1900, with much historical matter. See 1248.

881 Blanchard, Julian, "The History of Electrical Resonance." B. S. T. J., 20:415-433, Oct. 1941. 6 diags. Tuned circuits, discoveries and applications in a-c circuits, telephony, radio. A survey up to 1900. 32 refs. from 1826.

882 Ramsay, John F., "Microwave Antenna and Waveguide Techniques Before 1900." Proc. IRE., 46:405-415, Feb. 1958. 15 illus., 27 footnote refs. from 1889. A survey of quasi-optical experiments and devices from 1888 to 1900.

B. Developments from 1901

883 MacDonald, H.M., Electric Waves: Being an Adams Prize Essay in the University of Cambridge. Cambridge: University Press, 1902. 200 pp. On Faraday's and Maxwell's theories, and the characteristics of propagation, radiation, and transmission media.

884 Fleming, John A., "Electric Waves." The Principles of Electric Wave Telegraphy. Pt. 2, Chaps. 4-6, pp. 239-418; Stationary electric waves on wires, electromagnetic waves, detection and measurement of electric waves. 2nd ed., pp. 295-509, same with expanded treatment, also later eds. See 1253.

885 Braun, Karl F., "Electrical Oscillations and Wireless Telegraphy." Nobel lecture, Dec. 11, 1909. See 213, pp. 226-245. Biography, pp. 246, 247. Presentation speech, pp. 193-195. 19 photos and diags. (With G. Marconi, see 1259.) See also 389.

886 Lodge, Oliver J., The Ether of Space. London; New York: Harper and Brothers, 1909. 168 pp., 19 illus. The author, a foremost supporter of the ether theory, describes his experiments and gives some historical background from the time of Isaac Newton. Some refs. in the text. No index.

887 Pierce, George W., "Maxwell's Theory." Principles of Wireless Telegraphy, Chap. 7, pp. 36-41. Also following chaps., pp. 42-139, on the experiments of Hertz, electric waves and light, propagation on wires, wireless telegraphy before Hertz, Marconi's experiments, resonant circuits, nature of the oscillation, grounding, and propagation over the earth. See 1260.

888 Thomson, Joseph J., "Electric Waves." Encyclopaedia Britannica, 1911. 11th ed., 9:203-208. 10 diags. Numerous refs. in the text.

889 Zenneck, Jonathan, "The Antenna." Wireless Telegraphy, Chap. 6, pp. 150-172. Also, Chap. 10, "Propagation of the waves over the earth's surface," pp. 246-271. Other parts of the book are also pertinent. See 1272.

890 Eccles, William H., "Antennae and Earths." Wireless Telegraphy and Telephony, pp. 116-147. Also "Propagation of Waves," pp. 148-185. Historical matter and refs. in the text. See 1276.

891 Burrows, Charles R., "The History of Radio Wave Propagation Up to the End of World War I." Proc. IRE., 50:682-684, May 1962. 37 refs. from 1864.

892 Carter, Philip S., and Harold H. Beverage, "Early History of the Antennas and Propagation Field Until the End of World War I, Part I--Antennas." Proc. IRE., 50:679-682, May 1962. 14 footnote refs. from 1908.

893 Carson, John R., "Notes on the Theory of Modulation." Proc. IRE., 10:57-64, Feb. 1922. Reprinted Proc. IEEE., 51:893-896, June 1963. 4 footnotes. Mathematical analysis of frequency modulation.

894 Marconi, Guglielmo, "Radio Telegraphy." Proc. IRE., 10:215-238, Aug. 1922. 12 photos and diags. Joint AIEE-IRE lecture in New York, June 20, 1922. Survey of early long-distance communications, recent developments in multiple-tube transmitters, receiver design, propagation observations, point-to-point transmission with reflectors on short waves (1 to 20 meters). A pioneer paper.

895 Beverage, Harold H., Chester W. Rice, and Edward W. Kellogg, "The Wave Antenna. A New Type of Highly Directive Antenna." Trans. AIEE., 42:215-266, Feb. 1923. 85 illus., 12 tables, 26 footnotes from 1896. Bibliography, pp. 263-266, 172 items from 1894.

896 Lodge, Oliver J., Ether and Reality. London: Hodder and Stoughton, 1925. 179 pp. "A series of discussions on the many functions of the ether of space."

897 Leutz, Charles R., and Robert B. Gable, Short Waves. Altoona, Pa.: C.R. Leutz, 1930. 384 pp., 186 photos, maps, charts, diags. Chap. 1, pp. 11-32, Historical review. Other chapters on propagation, commercial and ship to shore radio, directional antennas, aircraft radio,

broadcast receivers, medical applications, amateur radio, and television. No refs.

898 [N. B. S.]. "Bibliography on Radio Wave Phenomena and Measurement of Radio Field Intensity." Proc. IRE., 19:1034-1089, June 1931. Prepared by the Bureau of Standards. 620 items from 1900, arranged chronologically under 14 headings. A list of journals and an author index are included. Comprehensive.

899 Free, E. E., "Searchlight Radio with the New 7-inch Waves." Radio News, 13:107-109, 152, Aug. 1931. 10 illus. Brief account of the English Channel demonstration. See also 1025.

900 Marconi, Guglielmo, "Micro Radio Waves." Electrician, 110:3-6, Jan. 6, 1933. 3 photos, 5 diags. Recent experiments on very short waves, with illustrations of transmitting equipment, circuits and wave patterns. Part of an address given at the Royal Institution, Dec. 2, 1932. Also 109:758, Dec. 9, 1932.

901 Armstrong, Edwin H., "The Spirit of Discovery. An Appreciation of the Work of Marconi." Elect. Eng., 72:670-676, Aug. 1953. A study of Marconi's three great discoveries in the field of radio transmission and reception. 8 refs. from 1898. Also a condensed version in Tele-Tech & Electronic Industries, 12:57, 222, Aug. 1953. A guest editorial by Armstrong: "The 'Bending' of the Microwaves." Marconi's third great discovery, his prophecy of 1932.

902 Armstrong, Edwin H., " 'Frequency' vs. 'Amplitude' Modulation." Radio-Craft, 7:75, 102, Aug. 1935. 2 photos of experimental equipment. Brief description of the new wideband frequency modulation system.

903 Armstrong, Edwin H., "A Method of Reducing Disturbances in Radio Signaling by a System of Frequency Modulation." Proc. IRE., 24:689-740, May 1936. 37 illus. A pioneer paper presented in New York on November 6, 1935. 13 refs. from 1912.

904 Armstrong, Edwin H., "Evolution of Frequency Modulation." Elect. Eng., 59:485-493, Dec. 1940. 14 illus. A survey of the history of FM with discussion of its advantages and applications. 4 refs.

905 Chu, Lan J., "Growth of the Antennas and Propagation Field Between World War I and World War II, Part 1--Antennas." Proc. IRE., 50:685-687, May 1962. 24 refs. from 1921.

906 Attwood, Stephen S., "Radio-Wave Propagation Between World Wars I and II." Proc. IRE., 50:688-691, May 1962. 36 refs. in 22 entries.

907 Armstrong, Edwin H., "Frequency Modulation and Its Future Uses." Ann. Am. Acad., 213:153-161, Jan. 1941. A nontechnical survey by the inventor.

908 Dix, J.C., "Early Centimetre-Wave Communication Systems." Engineering, 163:489, June 13, 1947. Brief summary of developments to 1941.

909 Rettenmeyer, Francis X., "Radio-Electronic Bibliography." Radio, pp. 35-38, June 1942. 2, Frequency Modulation, 210 items from 1922. See 93.

910 Rettenmeyer, Francis X., "Radio-Electronic Bibliography." Radio, pp. 36-38, 40, 42; 42, 44, 46, 48, 50, Feb., March 1943. 8, Antennas and Radiation, 542 items from 1900. See 93.

911 Rettenmeyer, Francis X., "Radio-Electronic Bibliography." Radio, pp. 35-40, April 1943. 9, Propagation, 231 items from 1900. See 93.

912 Kemp, J., "Waveguides in Electrical Communication." J. IEE., 9(Pt. 3):90-114, Sept. 1943. Historical review, 52 refs., including patents.

913 Perrine, J.O., "Electric Waves--Long and Short." Sci. Mon., 58:33-41, Jan. 1944. 3 illus. Historical survey.

914 Rettenmeyer, Francis X., "Radio-Electronic Bibliography." Radio, 28:38, 40, 42, Jan. 1944. 14, Interference and Static, 165 items from 1924. See 93.

915 Van Atta, Lester C., and Samuel Silver, "Contributions to the Antenna Field During World War II." Proc. IRE., 50:692-697, May 1962. 8 illus.

916 Norton, Kenneth A., "Radio-Wave Propagation During

World War II." Proc. IRE., 50:698-704, May 1962. Table, 62 refs.

917 Windlund, E. S., "Survey of Radio-Frequency Transmission Lines and Waveguides." Proc. Radio Club of America, 28(No. 2):1-64, 1951. A survey of the literature from 1919 to 1936, with 685 entries from 1919 to 1951.

918 [I. R. E.]. "Tropospheric Propagation: A Selected Guide to the Literature." Proc. IRE., 41:588-594, May 1953. Prepared by the IRE Committee on Wave Propagation. Brief survey arranged in 12 sections. 41 footnote refs. from 1918.

919 Benton, Mildred C. (comp.), Single-Sidebands in Communication Systems: a Bibliography. Washington: Office of Technical Services, 1956. U. S. Naval Research Laboratory, Bibliography No. 9. 99 pp., 492 entries from 1920. Papers, books and reports, with author and subject indexes.

920 Pierce, John R., Electrons, Waves and Messages. Garden City, N. Y.: Hanover House, 1956. 318 pp., 98 diags. Introduction for the general reader. Electric and magnetic fields, waves, Maxwell's equations, signals, noise, radiation, communication theory, microwaves. Slight historical background.

921 Oswald, Arthur A., "Early History of Single-Sideband Transmission." Proc. IRE., 44:1676-1679, Dec. 1956. 58 refs. from 1907.

922 Button, Kenneth J., "Historical Sketch of Ferrites and Their Microwave Applications." Microwave J., 3:73-79, March 1960. 5 graphs. Bibliography, 82 items from 1890, mainly of the 1950s.

923 Smith-Rose, Reginald L., "Fifty Years Research in Radio Wave Propagation." Wireless World, 67:203-207, April 1961. Diags.

924 Allen, Edward W., and Herman Garlan, "Evolution of Regulatory Standards of Interference." Proc. IRE., 50:1306-1311, May 1962. Primarily American practices. 45 refs. from 1928.

925 Southworth, George C., "Survey and History of the

Progress of the Microwave Arts." Proc. IRE., 50: 1199-1206, May 1962. 10 illus., 26 refs. from 1893. See next entry.

926 Southworth, George C., Forty Years of Radio Research. New York: Gordon and Breach, 1962. 274 pp., 59 illus. A "reportorial account" that is partly autobiographical. The author, for many years with Bell Telephone Laboratories, describes early radio (up to page 124). In the remainder of the book he describes the development of waveguides and his own pioneering contributions in this field. Some refs. in footnotes and in the text.

927 Heising, Raymond A., "Modulation Methods." Proc. IRE., 50:896-901, May 1962. 8 footnotes. Brief review of early radio telephone experiments and developments, including FM, from 1902.

928 Beverage, Harold H., "Antennas and Transmission Lines." Proc. IRE., 50:879-884, May 1962. Highlights in development from the turn of the century. 40 refs. from 1898.

929 Jordan, Edward C., and Ronold W. P. King, "Advances in the Field of Antennas and Propagation Since World War II: Part 1--Antennas." Proc. IRE., 50:705-708, May 1962. 9 illus.

930 Manning, Laurence A., "Radio Propagation Following World War II." Proc. IRE., 50:709-711, May 1962. Review of advances in new propagation modes. 17 refs.

931 Vogelman, Joseph H., "Microwave Communications." Proc. IRE., 50:907-911, May 1962. Brief historical summary from 1864, with a few names and dates in the text.

932 Deloraine, E. Maurice, and Alec H. Reeves, "The 25th Anniversary of Pulse Code Modulation." IEEE Spectrum, 2:56-63, May 1965. 2 photos. Historical background, by E.M. Deloraine; The past, present and future of PCM, by A.H. Reeves, 39 refs. from 1937, including patents.

933 Gunther, Frank A., "Tropospheric Scatter Communications. Past, Present, and Future." IEEE Spectrum, 3:79-100, Sept. 1966. 9 diags., 1 photo, 2 maps

and table of systems and companies, including a triple-fold display. 20 refs. from 1933.

934 Greenwood, Thomas L., "The Radio Spectrum Below 550 kHz." IEEE Spectrum, 4:121-123, March 1967. This brief survey includes much historical matter.

935 Schweitzer, Ellis, "40 Years of Waveguides; A Glimpse at History." Bell Lab. Rec., 48:72-79, March 1970. A pictorial essay with 18 photos and supporting text.

C. Radio Astronomy

936 Jansky, Karl G., "Directional Studies of Atmospherics at High Frequencies." Proc. IRE., 20:1920-1932, Dec. 1932. Jansky's pioneer paper on short-wave static. 14 photos, graphs, diags. Graphs show the direction of arrival and the intensity of static received on a rotating directional antenna. 6 refs. from 1926.

937 Jansky, Karl G., "Electrical Disturbances Apparently of Extraterrestrial Origin." Proc. IRE., 21:1387-1398, Oct. 1933. 9 graphs. 6 footnotes.

938 Jansky, Karl G., "Electrical Phenomena that Apparently are of Interstellar Origin." Popular Astronomy, 41:548-555, Dec. 1933.

939 Jansky, C.M., "The Beginnings of Radio Astronomy." American Scientist, 45:5-12, Jan. 1957. Photo.

940 Southworth, George C., "Early History of Radio Astronomy." Sci. Mon., 82:55-66, Feb. 1956. 12 illus. A review of Jansky's work on static, with references to his notebooks and reports; and of the author's early work on microwaves and studies of radio waves from the sun up to 1945. 8 refs.

941 Reber, Grote, "Early Radio Astronomy at Wheaton, Illinois." Proc. IRE., 46:15-23, Jan. 1958. A recounting of the author's pioneer experiments from 1936 to 1947. Design details of the parabolic antenna and various receivers are mentioned, along with some of the experimental observations. 23 refs. from 1932.

966 Shiers, George, "The First Electron Tube." Sci. Am., 220:104-112, March 1969. 11 illus., including a double-page chart showing the chronology of invention from 1879 to 1907. Experiments on thermionic emission and the contributions of Edison, Preece, Hittorf, Goldstein, Elster and Geitel, and Fleming that led to the latter's use of the diode as a wireless detector in 1904, and to the de Forest "Audion" of 1907.

967 Shunaman, Fred, "What Did de Forest Really Invent?" Radio-Electronics, 22:47-49, Sept. 1961. 7 illus.

968 Chipman, Robert A., "De Forest and the Triode Detector." Sci. Am., 212:92-100, March 1965. Port., 8 diags. Pioneering background and early radio detectors; coherer, electrolytic, diode, triode. A critical review of Lee de Forest's invention, his patent claims and his understanding of the "trigger action" of the three-electrode tube.

969 Wilkerson, D. C., "A Triode Antedating De Forest's Tube?" Radio News, 6:1671, 1754, 1756, March 1925. 4 diags. On the von Lieben tube.

970 Armstrong, Edwin H., "Operating Features of the Audion." Electrical World, 64:1149-1152, Dec. 12, 1914. 15 illus., circuits, graphs and photos. The first disclosure to correctly describe how a three-electrode tube operates as a detector and as a high-frequency amplifier. Graphs show a characteristic curve and current and voltage waveforms. Circuit connections for oscillographic recording and four photos of waveform traces. See also 1344.

971 Langmuir, Irving, "The Pure Electron Discharge and Its Applications in Radio Telegraphy and Telephony." Proc. IRE., 3:261-293, Sept. 1915. 10 illus. Discussion, pp. 287-293, 3 illus. A pioneer paper on high-vacuum tubes, techniques and applications, with historical background. Refs. in the text.

972 Eccles, William H., "Ionic Tubes." Wireless Telegraphy and Telephony, pp. 240-259. Also, Thermionic detectors, pp. 288-305. Tube developments, circuits, patents and applications up to about 1915. Names and patent numbers in the text, but no complete refs. See 1276.

973 de Forest, Lee, "The History of the Radio Tube 1900-1916." Radio News, 24:8, 55, 56, Dec. 1940. 2 illus.

974 Bucher, Elmer E., Vacuum Tubes in Wireless Communication. A Practical Text Book for Operators and Experimenters. New York: Wireless Press, 1918. 202 pp., 148 photos, graphs and diags. This pioneer text describes basic tubes with 2, 3, or 4 elements and discusses practical applications; amplification, regeneration, detection, modulation. Full coverage of contemporary practices, many circuit diagrams, but few specific references.

975 Smith-Rose, Reginald L., "The Evolution of the Thermionic Valve." J. IEE., 56:253-266, April 1918. 16 diags. Also, Wireless World, 6:10-17, 78-84, 142-147, 229-231, 281-283, April-Aug. 1918. Thorough survey that includes elementary thermionics, Fleming oscillation valve, de Forest "Audion" detector and amplifier, Lieben-Reisz valve, the pure electron discharge, and rectifier, amplifier and generator uses. 34 refs. from 1873.

976 Coursey, Philip R., "Vacuum Oscillation Generators." Telephony Without Wires, Chap. 13, pp. 161-192. Overview of developments to the end of World War I. 102 chap. refs. from 1890 include patents. See 1283.

977 Fleming, John A., The Thermionic Valve and Its Developments in Radiotelegraphy and Telephony. London: Wireless Press, 1919. 279 pp., 144 illus. App., pp. 256-274, text of the judgment in the U.S. Circuit Court, Sept. 1916. This important book provides a thoroughly documented record of the history of the electron tube applied to the radio arts. Historical introduction, pp. 1-45; The Fleming oscillation valve, pp. 46-101; The three-electrode valve, pp. 102-147; The thermionic valve as a generator of electric oscillations, pp. 148-175; The thermionic detector in radiotelegraphy, pp. 176-209; The thermionic oscillator and detector in radiotelephony, pp. 210-221; Recent improvements in thermionic apparatus, pp. 222-255. 59 footnotes from 1873, mostly refs., some multiple entries, with numerous patent numbers. Many other refs. and sources in the text. 2nd ed., 1924, 438 pp.

978 McNicol, Donald, "The Magic Bulb." Radio's Conquest of Space, Chap. 15, pp. 156-168. Also, Chap. 16, pp. 169-184, Vacuum tube development. Early history up to the end of World War I. 15 refs. in footnotes, other names and dates in the text. Following chapters also contain matter relevant to electron tube developments. See 1211.

979 Scott-Taggart, John, Thermionic Tubes in Radio Telegraphy and Telephony. London: Wireless Press, (1921). 424 pp., 344 illus. An exhaustive treatment of radio techniques as well as radio tubes. Coverage includes theory of thermionic currents, two-electrode and three-electrode tubes, detectors, amplifiers, continuous-wave transmitters and receivers, measuring instruments, and wireless telephony. The dynatron and other special tubes are also mentioned. Descriptive and partly tutorial, this book documents the progress of radio tube technology during and just after World War I. Over 100 footnote refs. to original articles and patents, many more refs. in the text.

980 Hawks, Ellison, "Fleming Discovers the Thermionic Valve; de Forest adds the Third Electrode." Pioneers of Wireless, Chap. 18, pp. 261-284. Edison effect, developments and applications by Fleming, de Forest, von Lieben, Wehnelt, Coolidge, Langmuir, Housekeeper, Hull. 12 footnotes. See 369.

981 de Forest, Lee, "The Audion--Its Action and Some Recent Applications." Radio News, 2:208-210; 280-282, 333, 335; 358, 359, 386, 388, Oct., Nov., Dec. 1920. 25 illus. Primarily historical.

982 Van der Bijl, H. J., The Thermionic Vacuum Tube. New York: McGraw-Hill, 1920. 391 pp., 232 illus. A pioneer book on the physics and characteristics of thermionic tubes and their uses as rectifiers, amplifiers, oscillators, modulators, detectors, and for miscellaneous purposes. 123 footnotes, mainly refs. to sources.

983 Richardson, Owen W., and W. Wilson, "Thermionics, Discharge of Electricity from Hot Bodies." Dictionary of Applied Physics, Vol. 2, pp. 868-882. 16 graphs and diags. Theoretical and experimental, with some historical matter. 35 footnotes, mainly refs. See 51.

984 Smith-Rose, Reginald L., "Thermionic Valves, Their Use in Radio Measurements." Dictionary of Applied Physics, Vol. 2, pp. 893-904. 13 diags. Methods of determining characteristics, application of triodes to radio measurements, oscillation generators. 64 refs., mainly 1916 to 1920. See 51.

C. Progress from 1921

985 Sharp, Clayton H., "The Edison Effect and Its Modern Applications." J. AIEE., 41:68-78, Jan. 1922. 16 photos, graphs, diags. Historical survey of electron-tube developments; diodes, triodes, power rectifiers, X-ray tubes.

986 Gilbert, Charles, "The True Story of the de Forest Vacuum Tube." Radio News, 4:633, 692, Oct. 1922. 2 illus.

987 Imlach, George, "The Four Electrode Vacuum Tube. Its Theory and Operation." Radio News, 6:694, 695, 740, Nov. 1924. 8 circuit diags., photo.

988 Gossling, B. S., and M. Thompson, "The Development of Valves for Wireless." World Power, 3:195-203, 333-339; 4:147-154, April, June, Sept. 1925. 27 refs.

989 Hull, Albert W., and N. H. Williams, "Characteristics of Shielded-Grid Pliotrons." Phys. Rev., 27:432-438, April 1926. 8 illus., 5 footnotes. Also, Hull, A. W., "Measurements of High Frequency Amplification with Shielded-Grid Pliotrons." Phys. Rev., 27:439-454, April 1926. 6 illus., 1 table, 14 footnotes. Basic papers on the tetrode.

990 Round, Henry J., The Shielded Four-Electrode Valve. London: Cassell, 1927. 87 pp. Capt. Round developed the screen-grid valve for the Marconi Company.

991 Goddard, F., The Four-Electrode Valve. London: Mills and Boon, 1927. 105 pp.

992 de Forest, Lee, "Evolution of the Vacuum Tube." Radio News, 11:990, 991, 1039, May 1930. 8 illus.

993 de Forest, Lee, "How the Radio Tube Grew Up." Radio News, 13:23, 74, 75, July 1931. 4 illus. On electron-tube developments during the early 1900s and the manufacture of radio tubes.

994 O'Dea, William T., "The Thermionic Valve." Radio Communication: Its History and Development, pp. 38-49. Soft valve, hard valve, screening electrodes, indirectly heated cathodes, multiple valves, "Class B" valves, re-

ceiving valves, transmitting valves, construction, demountable valves. A concise history to 1931. See 1206.

995 Chaffee, E. Leon, Theory of Thermionic Vacuum Tubes. New York: McGraw-Hill, 1933. 652 pp., illus. Standard text. Over 400 refs.

996 Koller, L. R., The Physics of Electron Tubes. New York: McGraw-Hill, 1934. 205 pp., 67 photos, graphs, diags., 14 tables. Emphasizes physical phenomena; emission, emitters, cathodes, secondary emission, gases, space charge, discharge in gases, photoelectric effects, with some manufacturing details. Chap. refs., 108 items, mainly of the 1920s.

997 Mullard, S. R., "The Development of the Receiving Valve." J. IEE., 76:10-16, 1935. 20 illus., including 1 plate of photos. General survey from 1922.

998 Fink, Donald G., Engineering Electronics, New York: McGraw-Hill, 1938. 358 pp., 217 illus. Pt. 2, pp. 93-221, Electron tubes. Pt. 3, pp. 225-331, Electron-tube applications. Bibliography, 86 items.

999 MacArthur, Elmer D., "An Indexed Bibliography of Electron Tubes and Their Applications." G. E. Rev., 41:455-460, Oct. 1938. 138 entries from 1921.

1000 Ingram, S. B., "A Decade of Progress in the Use of Electronic Tubes. Part 1--In the Field of Communication." Elect. Eng., 59:643-649, Dec. 1940. Bibliography of 88 items referenced in the text. Classifications are: Review articles on communication, broad-band carrier systems, ultrashort-wave tubes, beam power tubes, wide-band amplifier tubes, secondary-emission multipliers, television and television tubes, gas tubes, frequency modulation, ultrashortwave circuits.

1001 Electron Tubes. Volume 1, 1935-1941. Princeton, N. J.: RCA Review, 1949. 475 pp. Reprints of 19 papers grouped under general, transmitting, receiving, and special headings, with 32 summaries of other papers. App. 1, pp. 455-471, Bibliography of technical papers by RCA authors from 1919 to 1941, over 400 entries arranged by year. App. 2, pp. 472-475, List of application notes, 1933-1941, 117 items.

1002 Rettenmeyer, Francis X., "Radio-Electronic Bibliography." Radio, pp. 39, 40, 42, 44, 46, Aug. 1942. 4, Tubes, 219 items from 1915. See 93.

1003 White, William C., "Electron Tube Terminology." Electronics, 15:42-45, 154, Dec. 1942. 2 illus., 2 tables. History of electron tube names and classification of tubes. 6 refs.

1004 Mouromtseff, Ilia E., "A Quarter Century of Electronics." Elect. Eng., 66:171-177, Feb. 1947. 5 illus. A brief outline of the development of electron tubes, including manufacturing. 38 refs. from 1922.

1005 de Forest, Lee, "The Audion." Elect. Eng., 66:255-257, March 1947. 2 photos. Remarks on receiving the 1946 Edison Medal. Also, Introduction by David Sarnoff, pp. 254, 255.

1006 Gorham, John E., "Electron Tubes in World War II." Proc. IRE., 35:295-301, March 1947. 8 photos, 2 tables. Survey of advances in electron tubes; general research, magnetrons, transmit-receive tubes, klystrons, planar tubes, indicator and pickup tubes, power and gas tubes, receiving tubes, and crystal rectifiers.

1007 Electron Tubes. Volume 2, 1942-1948. Princeton, N.J.: RCA Review, 1949. 454 pp. Reprints of 21 papers grouped under general, transmitting, receiving, and special headings, with 20 summaries of other papers. App. 1, pp. 444-453, Bibliography of technical papers by RCA authors from 1942 to 1948, nearly 200 entries arranged by year. App. 2, List of application notes, 1947, 1948, (118-137, continued from Vol. 1).

1008 Fink, Donald G., "Electron Tubes 1930 to 1950." Electronics, 23:66-69, April 1950. Chronological list of new tube types, 3 tables.

1009 Howe, G. W. O., "Genesis of the Thermionic Valve." Engineer, 198:745, 746, Nov. 26, 1954.

1010 Watson, Paul G., "The Electron Tube." Radio and Television News, 52:66, 67, 166, 167, Nov. 1954. 6 photos, 3 show collections of early tubes with identifying captions.

1011 Report on the Supply of Electronic Valves and Cathode Ray Tubes. (The Monopolies and Restrictive Practices Commission.) London: H.M.S.O., Dec. 1956. Pages 9-13, historical references to manufacturers.

1012 Herold, Edward W., "The Impact of Receiving Tubes on Broadcast and TV Receivers." Proc. IRE., 50: 805-810, May 1962. A short but thorough survey of tube and circuit developments up to 1960 with numerous references to tube types and dates in the text. 59 refs. from 1905.

1013 Johnson, John B., "Contributions of Thomas A. Edison to Thermionics." Am. J. Phys., 28:763-773, Dec. 1960. 9 illus., including pages from Edison's notebooks. 7 refs. from 1884. See also 959.

D. Industrial Electron Tubes

1014 Hull, Albert W., "Gas-filled Thermionic Tubes." J. AIEE., 47:798-803, Nov. 1928. 8 illus., 6 refs. Also Trans. AIEE., 47:753-763, 1928. On thermionic tubes with low-pressure gas, their characteristics as rectifiers and thyratrons.

1015 King, W.R., "Electron Tubes in Industry." Trans. AIEE., 50:590-598, June 1931. Photoelectric tubes, thyratrons; characteristics and applications as relays, controlled rectifiers, color analysis and matching.

1016 Hull, Albert W., "Characteristics and Functions of Thyratrons." Physics, 4:66-75, Feb. 1933. 16 illus., 16 footnote refs. from 1924.

1017 Henney, Keith, Electron Tubes in Industry. New York: McGraw-Hill, 1934, 490 pp. 2nd ed., 1937, 539 pp., about 400 illus. This extensive text deals with thermionic tubes, amplifiers, gaseous triodes, light-sensitive tubes and applications, rectifiers, cathode-ray tubes and miscellaneous tubes and circuits. More than 60 items in chapter bibliographies, other refs. in the text, numerous footnotes.

1018 Fleming, John A., "The Thermionic Valve in Scientific Research." J.F.I., 220:151-165, Aug. 1935. 3 diags. Early history is followed by a discussion of special tubes, and of electron-tube applications for measurements. Some refs. in the text. Also, Science, 81:625-628, June 28, 1935.

1019 White, William C., "A Decade of Progress in the Use of Electronic Tubes. Part II--In Other Than the Field of Communication." Elect. Eng., 59:650-654, Dec. 1940. A summary of tube types and applications; power and industrial. Bibliography has 144 entries in classified groups, "chosen by careful elimination from several thousand."

1020 Cobine, James D., Gaseous Conductors. Theory and Engineering Applications. New York: McGraw-Hill, 1941. Reprinted, Dover, 1958. 606 pp., 350 photos, graphs, diags., 61 tables. Physical concepts of gaseous conduction, discharge characteristics, engineering applications in circuit interrupters, rectifiers, light sources, oscillographs. General refs., p. 593, 21 items; 650 footnotes, mainly refs., many multiple entries. Article titles not given.

1021 Cobine, James D., "The Development of Gas Discharge Tubes." Proc. IRE., 50:970-978, May 1962. 6 illus. Brief but full review of ionization, cold-cathode, hot-cathode arc, liquid metal arc, and plasma tubes. 107 refs. from 1785.

E. Microwave Tubes

1022 Hull, Albert W., "The Magnetron." J. AIEE., 40: 715-723, Sept. 1921. 29 photos, graphs, diags. Description, theory, applications. A lecture given in New York on May 20, 1921.

1023 Hull, Albert W., "The Axially Controlled Magnetron. A New Type of Magnetron, Controlled by Current Through the Filament." Trans. AIEE., 42:915-920, discussion p. 953, June 1923. 11 illus., table, 4 refs. from 1916.

1024 White, William C., "Some Events in the Early History of the Oscillating Magnetron." J. F. I., 254:197-204, Sept. 1952. 6 illus. Primarily about the early work of Hidetsugu Yagi and Kinjiro Okabe up to 1930. 16 refs. from 1921.

1025 Saxl, Irving J., "The World's Highest Frequency

Oscillator Tube." Radio News, 13:265-267, Oct. 1931. 8 illus. First description of the "micro-ray" tube used in the Dover-Calais microwave demonstration. See also 899.

1026 Varian, Russell H., and Sigurd F. Varian, "A High Frequency Oscillator and Amplifier." J. App. Phys., 10:321-327, May 1939. 6 illus., 3 refs. First article by the Varian brothers on the klystron. An expansion of a letter, p. 140, Feb. 1939.

1027 Varian, Russell H., "The Invention and Development of the Klystron." Military Automation, Sept.-Oct. 1957, pp. 256-259. 5 illus. A personal account of the circumstances that led to the first operable tube.

1028 Randall, John T., and Henry A. H. Boot, "Early Work on the Cavity Magnetron." J. IEE., 93(Pt. 3A):182, 183, March 1946. Diag. On original work in the Physics Department of the University of Birmingham in Nov. 1939.

1029 Harvey, Arthur F., Thermionic Tubes at Very High Frequencies. London: Chapman and Hall; New York: John Wiley, 1941. 244 pp. 2nd ed., High Frequency Thermionic Tubes, 1943. 235 pp., 99 photos, charts and diags. General properties, influence of frequency, retarding field generators, the magnetron, miscellaneous tubes and circuits, wave guides and horn radiators. Chap. refs., 517 entries.

1030 Boot, Henry A. H., and John T. Randall, "The Cavity Magnetron." J. IEE., 93(Pt. 3A):928-938, March 1946. An account by the inventors.

1031 Randall, John T., "The Cavity Magnetron." Proc. Phys. Soc., 58:247-252, May 1946. 6 illus. On early work in England from late 1939 to 1942. 9 refs. from 1897.

1032 Kompfner, Rudolf, The Invention of the Traveling-Wave Tube. San Francisco: San Francisco Press, 1964. 30 pp., 18 diags. A first-hand story giving step-by-step details of the process of invention as well as of the invention itself. Up to the mid-1940s. 21 refs. from 1931.

1033 Wathen, Robert L., "Genesis of a Generator--The

Early History of the Magnetron." J. F. I., 255:271-287, April 1953. 8 illus. A survey from 1921 through English work during the early years of World War II. 38 refs. from 1924.

1034 Rettenmeyer, Francis X., "Radio-Electronic Bibliography." Radio, 29:52; 52, 53, 73, June, July 1945. 18, Velocity Modulation, 138 items from 1935. See 93.

1035 Megaw, Eric C. S., "The High-Power Pulsed Magnetron: A Review of Early Developments." J. IEE., 93(Pt. 3A):977-984, March 1946.

1036 Barton, Martin A., "Traveling Wave Tubes." Radio, 30:11-13, 30, 31, 32, Aug. 1946. 5 diags. Based on BTL information; said to be the first published discussion.

1037 Kompfner, Rudolf, "The Travelling Wave Valve." Wireless World, 52:369-372, Nov. 1946. The inventor's first paper.

1038 Suits, C. Guy, "Physics Today--Engineering Tomorrow." Elect. Eng., 66:241-243, March 1947. On the magnetron.

1039 Wathen, Robert L., "The Traveling Wave Tube--A Record of Its Early History." J. F. I., 258:429-442, Dec. 1954. 6 illus. On the development of theory and experimental results in England, France and the United States from 1935 to 1947. 29 refs. from 1924.

1040 Collins, George B. (ed.), Microwave Magnetrons. M. I. T. Rad. Lab. Series, Vol. 6, 1948. 806 pp., 526 illus. Theory, operation, design and construction. Chap. 1 includes a brief history of early types and of the British cavity magnetron (10 pp.). 9 illus., 13 refs. from 1921. See 798.

1041 Hamilton, Donald R., and others, Klystrons and Microwave Triodes. M. I. T. Rad. Lab. Series, Vol. 7, 1948. 533 pp., 226 illus. Fundamentals, planar space-charge tubes, klystrons. Numerous references to wartime reports, but no historical treatment. See 798.

1042 Kompfner, Rudolf, "Travelling-Wave Tubes." Reports on Progress in Physics, 15:275-327, 1952. 20 diags. and charts. Survey and development, mathematical treatment, with much historical background. Refs., pp. 325-327, 122 classified entries. 13 footnotes, other refs. in the text.

1043 Harvey, Arthur F., "Microwave Tubes: An Introductory Review with Bibliography." Proc. IEE., 107(Pt. C):29-59, March 1960. 29 illus. Pt. 1, Grid-controlled, space-charge wave and travelling-wave tubes. Pt. 2, Crossed-field tubes, ultramicrowave generation and electron sources. Pt. 3, Noise in electron tubes. Bibliography, pp. 51-59, 698 items arranged alphabetically by author for each part.

1044 Pierce, John R., "History of the Microwave-Tube Art." Proc. IRE., 50:978-984, May 1962. 5 tables. A short survey from 1931. 74 refs. from 1920.

1045 Kompfner, Rudolf, "The Development of the Travelling Wave Tube." Endeavour, 24:106-110, May 1965. 4 diags. 12 refs. from 1940.

11. INDUSTRIAL ELECTRONICS

Topics in this chapter include radio-frequency measurements, electronic instruments, components, assemblies, and computers. For other entries that treat related topics, see 250, 743, 767, 768, 794, 799, 808, 1017, also 1604.

A. General

1046 Gulliksen, Finn H., and Edwin H. Vedder, Industrial Electronics. London: Chapman and Hall; New York: John Wiley, 1935. 245 pp., diags. and photos. An early book on electron tubes, apparatus and appliances in industrial applications. Numerous chap. refs.

1047 White, William C., "Early History of Industrial Electronics." Proc. IRE., 50:1129-1135, May 1962. 6 illus. A general survey of rectifier tubes, early phototubes, high-voltage and power tubes, and high-frequency heating from 1912 to 1940. No refs.

1048 Batcher, Ralph R., and William Moulic, The Electronic Control Handbook. New York: Caldwell-Clements, 1946. 344 pp., 297 illus. Basic elements, conversion elements, activation elements, circuits, control motors. 110 chap. refs. Bibliography, pp. 336-338, 72 items.

1049 [Westinghouse Electric Corporation]. Industrial Electronics Reference Book. New York: John Wiley; London: Chapman and Hall, 1948. 680 pp., 1140 illus. This basic work by Westinghouse engineers covers the whole field in 36 chapters. Chap. refs., 895 items.

1050 Richter, Walther, "Industrial Electronic Developments in the Last Two Decades and a Glimpse into the Future." Proc. IRE., 50:1136-1142, May 1962. A general discussion with some names and dates in the text.

B. Instruments, Measurements, and Control

1051 Fleming, John A., "High Frequency Electric Measurements." The Principles of Electric Wave Telegraphy, Chap. 2, pp. 85-160. 2nd ed., pp. 111-211. 36 illus., 64 footnote refs. Full coverage of early techniques, instruments, methods, experiments and results. See 1253.

1052 [National Bureau of Standards]. Radio Instruments and Measurements. Washington: G. P. O., 1918. Circular 74, 318 pp., 224 illus. A pioneer text on the fundamentals, techniques, methods and instrumentation for radio-frequency measurements. Bibliography. Revised ed., 1924, frequent reprintings.

1053 Dye, D. W., "Radio-Frequency Measurements." A Dictionary of Applied Physics, Vol. 2, pp. 627-679. 95 photos, graphs, diags. 13 footnote refs. Extensive coverage of wave-length and frequency, current, crystal detector, condensers, inductance, resistance and decrement, cathode-ray tube. 162 refs. See 51.

1054 Moullin, Eric B., The Theory and Practice of Radio Frequency Measurements. London: Charles Griffin, 1926. 2nd ed., 1931. 487 pp., 284 illus., 45 tables. 300 footnotes, mostly refs. A pioneer text.

1055 Horton, J. W., "Use of Vacuum Tubes in Measurements." Elect. Eng., 54:93-102, Jan. 1935. Group I, Fundamental properties of vacuum tube circuits, vacuum tubes, other circuit elements, amplification, regeneration, modulation, harmonic generation, oscillation. Group II, Measuring methods and apparatus. General refs., 8 items. Bibliography, 596 entries.

1056 Rettenmeyer, Francis X., "Radio-Electronic Bibliography." Radio, pp. 29, 30, 32, 34, 36; 26, 29, 30, 32, 34; 39, 40, 42, 44, 46, Nov., Dec. 1942, Jan. 1943. 7, Remote control. 576 items, 117 patents with number, year, name and subject. See 93.

1057 Rettenmeyer, Francis X., "Radio-Electronic Bibliography." Radio, pp. 50, 52, 54, May 1943. 10, Measurements, General, 210 items from 1913. See 93.

1058 Stansel, F. R., "Bibliography on Frequency Measurements at R-F, including UHF." Radio, pp. 34, 36-38, 44, 46, 48, May 1943. 374 items from 1894.

1059 Rettenmeyer, Francis X., "Radio-Electronic Bibliography." Radio, 30:54, 56, 58, Aug. 1945. 19, Frequency Standards, 148 items. (This list prepared by the National Bureau of Standards.) See 93.

1060 Andres, Paul G., "Electronics in Instrumentation." Survey of Modern Electronics, Chap. 8, pp. 328-391. 45 photos and diags. 30 refs. from 1928. Also, Chap. 10, pp. 428-459, Electronic controls. 21 photos and diags. 25 refs. from 1933. Chap. 11, pp. 460-499, Electronics in heating. 28 photos and diags. 24 refs. from 1942. See 802.

1061 Greinacher, H., "The Evolution of Particle Counters." Endeavour, 13:190-197, Oct. 1954. 12 photos and diags. 16 refs. from 1908.

1062 Barbour, W. E., "Radioisotopes. A New Dimension for Applied Science." Elect. Eng., 78:423-434, May 1959. 13 photos. This survey of 25 years includes sensing devices, instruments and special components. 14 refs. from 1948.

1063 Sinclair, Donald B., "The Measuring Devices of Electronics." Proc. IRE., 50:1164-1170, May 1962. An overview of instruments with references to early papers from 1895.

1064 Oliver, Bernard M., "Digital Display of Measurements in Instrumentation." Proc. IRE., 50:1170-1172, May 1962. A short survey from the early 1940s. 12 refs. from 1944.

1065 Richardson, John M., and James F. Brockman, "The U.S. Basis of Electromagnetic Measurements." IEEE Spectrum, 1:129-138, Jan. 1964. 2 photos, 7 charts and graphs. This account of the work at the National Bureau of Standards contains some history of radio standards, especially in the chronological charts.

C. Electronic Components

1066 Fleming, John A., The Alternate Current Transformer. London: Benn Brothers, 1889, 1892, 2 vols. Vol. 2, Chap. 1, The historical development of the induction coil and transformer, provides a detailed description of coil design, construction, performance and improvements from the mid-1830s.

1067 Coursey, Philip R., Electrical Condensers: Their Construction, Design, and Industrial Uses. London: Pitman, 1927. 660 pp., 514 illus. This comprehensive work contains a bibliography with over 1500 items.

1068 Greenwood, Harold S., A Pictorial Album of Wireless and Radio, 1905-1928. Los Angeles: Floyd Clymer, 1961. 219 pp. Over 1100 pictures of components and apparatus, including 30 full-page reproductions of contemporary publications, brochures and advertisements. 32 groups with brief historical introductions. Individual items are identified by make, type, date and price. A unique record of a personal collection.

1069 Blackburn, John F. (ed.), Components Handbook. M.I.T. Rad. Lab. Series, Vol. 17, 1949. 626 pp., 369 illus. Mainly on the classes of components used or developed in the Radiation Laboratory for receiving and test equipment. Wires and cables, resistors, iron core inductors, piezoelectric devices, delay lines, potentiometers, special variable condensers, rotary inductors, motors and tachometers, power supplies, relays, receiving tubes. Over 100 refs., some grouped, others as footnotes. See 798.

1070 Dubilier, William, "Development, Design and Construction of Electrical Condensers." J.F.I., 248:193-204, Sept. 1949. 5 illus. The Leyden jar, mica, paper and electrolytic capacitors. On problems in design and construction, with some historical comments.

1071 Noltingk, B. E., "History of Electrical Devices for Measuring Strains and Small Movements." J. Sci. Insts., 35:157, 158, May 1955.

1072 Brothers, James T., "Historical Development of Component Parts Field." Proc. IRE., 50:912-919, May 1962. 2 diags. A general overview without refs.

1073 Marsten, Jesse, "Resistors--A Survey of the Evolution of the Field." Proc. IRE., 50:920-924, May 1962. 39 refs. from p. 884.

1074 Podolsky, Leon, "Capacitors." Proc. IRE., 50:924-928, May 1962. 4 tables. Primarily a chronological bibliography, with 98 entries from 1705 to 1913 in 24 categories.

D. Electronic Assemblies

1075 Sargrove, John A., "New Methods of Radio Production." J. Brit. IRE., 7:2-33, Jan.-Feb. 1947. Pioneer paper on printed circuits and mass production methods.

1076 Brunetti, Cledo, and Roger W. Curtiss, "Printed Circuit Techniques." Proc. IRE., 36:121-161, Jan. 1948. 51 charts, diags. and photos, 4 tables. This basic paper covers painting, spraying, chemical deposition, vacuum processes, die-stamping, dusting, performance and applications. 60 refs. Also published separately, N. B. S. Circular 465, Nov. 1947.

1077 Proctor, Warren G., and others, "The Design and Construction of Electronic Apparatus." Electronic Instruments, Chap. 19, pp. 667-708. 24 illus., 4 tables. On radar and navigational equipment built for service during World War II. See 1094.

1078 Barrett, R. M., "Microwave Printed Circuits, a Historical Survey." Trans. IRE., MTT-3:1-10, March 1955.

1079 Eisler, Paul, The Technology of Printed Circuits; the Foil Technique in Electronic Production. London: Heywood, New York: Academic Press, 1959. 405 pp., illus. Bibliography (comp., Mrs. K. Bourton) pp. 351-402, 480 items from 1945.

1080 Danko, Stanley F., and others, "The Micro-Module: A Logical Approach to Microminiaturization." Proc. IRE., 47:894-903, May 1959. 12 illus., table of construction systems. Bibliography, 6 items.

1081 Dummer, Geoffrey W. A., and others, Electronic Equipment Design and Construction. New York: McGraw-Hill, 1961. 241 pp., 84 illus. Chap. 2, pp. 4-17, A brief history and review of constructional methods. 102 chap. refs.

1082 Danko, Stanley F., "Printed Circuits and Microelectronics." Proc. IRE., 50:937-945, May 1962. 6 illus. A summary of developments since 1945. 44 refs. from 1947 in 5 groups.

1083 Robinson, Preston, "Electronic Materials, 1912-1962." Proc. IRE., 50:945-949, May 1962. An informal survey of dielectric, magnetic, ferroelectric and resistive materials, along with conductors, insulators and semiconductors. No refs.

1084 Batcher, Ralph R., and Alfred R. Gray, "The Influence of Product Design on Radio Progress." Proc. IRE., 50:1289-1305, May 1962. 19 illus. A survey of radio history, equipment, manufacturing progress and products from 1900 to 1961. Bibliography, 69 entries.

1085 Boehm, George A.W., "Electronics Goes Microminiature." Fortune, 66:98-102, 176-182, Aug. 1962. 7 illus. Thin-film circuits; research, development, computer applications and prospects.

1086 Parks, Martha S., The Story of Microelectronics. First, Second and Future Generations. Anaheim, Calif.: North American Aviation, Autonetics Division, 1966. 138 pp., nearly 300 illus., including color photos, charts, graphs, diags. A general survey with a light technical treatment. Some historical background is given in the first 25 pages. The bulk of the book (pp. 26-95) is devoted to company activities. There are 5 Apps.: How semiconductors and semiconductor devices work, Automated MOS mask preparation, Components reliability assurance, Available Autonetics technical publications on microelectronics (104 entries), Microelectronics glossary.

1087 Shiers, George, Design and Construction of Electronic Equipment. Englewood Cliffs, N.J.: Prentice-Hall, 1966. 362 pp., 230 illus. Bibliography, pp. 348-353, 110 items from 1954. Chap. 1, pp. 1-20, Historical development; early wireless equipment, early broadcast equipment, radio-electronics, modern electronics, electron tubes, components, apparatus and materials. A general survey, no refs.

E. Computers

1088 Eccles, William H., and F.W. Jordan, "A Trigger

Relay Utilizing Three-Electrode Thermionic Vacuum Tubes." Radio Amateur News, 1:349, Jan. 1920. 2 circuit diags. Description of the famous "flip-flop" circuit patented in 1918.

1089 Aiken, Howard H., "Proposed Automatic Calculating Machine." IEEE Spectrum, 1:62-69, Aug. 1964. Photo of the IBM automatic sequence-controlled calculator (Mark 1), dedicated August 7, 1944. This memorandum of 1937 includes historical background, need for new methods, mathematical operations and means, mechanical considerations and probable speed of computation. 5 refs. from 1946 are given in a "Note on the twentieth anniversary of the Mark 1 computer."

1090 "War Department Unveils 18,000-tube Robot Calculator." Electronics, 19:308-314, April 1946. 3 photos. An early news release on the ENIAC.

1091 Aiken, Howard H., and Grace M. Hopper, "The Automatic Sequence Controlled Calculator." Elect. Eng., 65:384-391, 449-454, 522-528, Aug.-Nov. 1946. Survey of mathematical processes and description of the mechanism, functional units and operation.

1092 Hartree, Douglas R., "The ENIAC, an Electronic Computing Machine." Nature, 158:500-506, Oct. 12, 1946.

1093 Tumbleson, Robert C., "Calculating Machines." Elect. Eng., 67:6-12, Jan. 1948. 3 photos, 2 diags., 1 table. Includes a little history along with description of the ENIAC, EDVAC, MANIAC and UNIVAC.

1094 Greenwood, Ivan A., and others (eds.), Electronic Instruments. M.I.T. Rad. Lab. Series, Vol. 21, 1948. 721 pp., 466 illus. Electronic analogue computers, instrument servomechanisms, voltage and current regulators, pulse test equipment. See 798.

1095 Brainerd, John G., and T.K. Sharpless, "The ENIAC." Elect. Eng., 67:163-172, Feb. 1948. 6 illus.

1096 Page, C.H., "Digital Computer Switching Circuits." Electronics, 21:110-118, Sept. 1948. 4 diags., 1 table. An early description of flip-flop and other circuits employing vacuum tubes for storage, register, memory and basic arithmetic functions.

1097 Davis, Harry M., "Mathematical Machines." Sci. Am., 180:28-39, April 1949. 11 photos. A general discussion of the historical development of calculating machines, with some speculations about future developments. Includes description of the ENIAC and the IBM selective sequence calculator.

1098 Hartree, Douglas R., Calculating Instruments and Machines. Urbana, Ill.: University of Illinois Press, 1949. 138 pp., 68 illus. Chap. 6, pp. 69-73, on Charles Babbage and the analytical engine. Chap. 7, pp. 74-93, covers the Harvard Mark 1, relay machines, the ENIAC and the IBM selective sequence electronic calculator. Refs., pp. 131-134, 122 items from 1842.

1099 Berkeley, Edmund C., Giant Brains or Machines That Think. New York: John Wiley; London: Chapman and Hall, 1949. 270 pp., 77 diags. An early work on computers written for the layman. Refs., pp. 228-255, 19 categories, 298 entries. Reprinted by Science Editions (New York), 1961. 294 pp. "Comments in 1961," pp. 256-279, are additional notes on each chapter.

1100 "Office Robots." Fortune, 45:82-87, 112, 114, 117, 118, Jan. 1952. 11 photos, one multiple diag. Business survey of electronic computers, types of machines, installations, manufacturers, applications and prospects. Some technical description with a little history.

1101 MacWilliams, W. H., "Computers--Past, Present, and Future." Elect. Eng., 72:116-121, Feb. 1953. Historical development of computers with comments on present problems and trends. 14 refs. from 1889.

1102 Booth, Andrew D., and Kathleen H. V. Booth, Automatic Digital Calculators. New York: Academic Press; London: Butterworth, 1953. 2nd ed., 1956. 261 pp. 605 refs. from 1842.

1103 Carroll, John M., "Electronic Computers for the Businessman." Electronics, 28:121-131, June 1955. 16 photos, 5 tables on cost, availability, technical characteristics, performance, speed, models. Bibliography, 35 items published in Electronics from 1946.

1104 Smith, Thomas M., "Origins of the Computer." Technology in Western Civilization, Vol. 2, pp. 309-323. 4 photos. Early mathematical machines, the work of Charles Babbage, Howard Aiken and the Harvard Mark I computer, the ENIAC, and electronic digital computers up to the mid-1950s. See 232.

1105 Watson, Thomas J., "Man's Most Versatile Machine." Elect. Eng., 78:500-508, May 1959. 8 photos. Historical background of data processing equipment and discussion of electronic computers, with special reference to IBM machines.

1106 Serrell, Robert, and others, "The Evolution of Computing Machines and Systems." Proc. IRE., 50:1039-1058, May 1962. 36 illus., 11 footnotes. Bibliography, classified by year from 1914, 142 entries. Brief survey of early machines; mechanical, electromechanical, and electronic.

1107 Halacy, D.S., Computers--The Machines We Think With. New York: Harper and Row, 1962. 279 pp., 78 illus. Chap. 2, pp. 18-47, The computer's past. Primarily on mechanical devices. A general survey.

1108 Bernstein, Jeremy, The Analytical Engine. Computers--Past, Present, and Future. New York: Random House, 1964. 113 pp. Bibliography, pp. 105-109, 15 items. A popular exposition based on articles in The New Yorker.

1109 Thomas, Shirley, Computers; Their History, Present Applications, and Future. New York: Holt, Rinehart and Winston, 1965. 174 pp., 26 illus. Chap. 3, pp. 53-75, History. 34 refs. Book list, 30 entries. Popular story.

1110 Swallow, K.P., and W.T. Price, Elements of Computer Programming. New York: Holt, Rinehart & Winston, 1966. Chap. 1, pp. 5-12, History of digital computers, 6 illus. A brief survey, primarily mechanical.

1111 Richards, R.K., Electronic Digital Systems. New York: John Wiley, 1966. 637 pp., illus. Chap. 1, pp. 1-49, History and introduction. Discusses electronic systems from the late 1930s, major computer prototypes, electromechanical machines, components and system concepts. The bibliography contains 10 refs. from 1952.

1112 Richards, R. K., Electronic Digital Components and Circuits. Princeton, N. J.: D. Van Nostrand, 1967. 526 pp., illus. Chap. 1, pp. 1-29, History and introduction. Circuit history from Eccles and Jordan (1919); diodes, transistors, integrated circuits, magnetic components, delay lines are briefly mentioned. No refs. Classed bibliographies (primarily article references from the mid-1950s) are furnished with all other chapters.

1113 Rosenberg, Jerry M., The Computer Prophets. New York: Macmillan, 1969. 192 pp., 16 plates. An introductory survey of computer developments from Pascal to the 1960s. Chapters are based on individuals and companies. Names and dates in the text, but no refs. Reading list, pp. 190-192, 36 items.

12. RADAR

Scientific work employing the reflection of radio waves during the 1920s was a prelude to techniques whereby aircraft could be "located" by radio means. Concurrent advances in shortwave radio, beam transmission, radio direction finding, directive antennas, high-frequency electron tubes, pulse circuits and related techniques, as well as progress in cathode-ray tubes and circuits, and especially television, all contributed to the rise of this new branch of electronic technology.

Apart from brief news releases in 1941 and 1943, the story of radar was not made public until 1945, long after its practical inception during the mid-1930s. The recorded history of radar is therefore largely a composite of official news releases, group histories and personal accounts published during and just after 1945 and later publications based upon prewar and wartime developments. See also 462, 537, 559, 566, 796-799, 810-812, 926, for related material.

A. Radio Prelude to Radar

1114 Taylor, A. Hoyt, "The Navy's Work on Short Waves." QST, 7:9-14, May 1924. 6 illus. On shortwave apparatus and proposed tests for the polar flight of the dirigible *Shenandoah*, with remarks on the behavior of short waves (5 meters), standing waves, reflections, effect of obstacles, daylight and night propagation.

1115 Breit, Gregory, and Merle A. Tuve, "A Radio Method of Estimating the Height of the Conducting Layer." Nature, 116:357, Sept. 5, 1925. A short note referring to the transmission of interrupted high-frequency wave-trains and the reception of double signals, one direct and the other reflected.

1116 Appleton, Edward V., and M. A. F. Barnett, "On Some Direct Evidence for Downward Atmospheric Reflection of Electric Rays." Proc. Roy. Soc., 109(Ser. A):621-641, Dec. 1925. On a frequency modulation method used in December 1924 to find the height of the Kennelly-Heaviside layer.

1117 Breit, Gregory, and Merle A. Tuve, "A Test of the Existence of the Conducting Layer." Phys. Rev., 28: 554-575, Sept. 1926. 11 illus., including circuits and photos of waveforms. Also, Sci. Am., 136:356, May 1927. Photo of authors and equipment.

1118 Taylor, A. Hoyt, and Leo C. Young, "Studies of High-Frequency Radio Wave Propagation." Proc. IRE., 16:561-578, May 1928. 7 illus. Also, "Studies of Echo Signals," 17:1491-1507, Sept. 1929. 11 illus. On multiple signals, echoes, directional characteristics, diurnal variations, frequency effects.

1119 Appleton, Edward V., and G. Builder, "A Simple Method of Investigating Wireless Echoes of Short Delay." Nature, 127:970, June 27, 1931. One illus. showing 3 waveforms. Ionospheric experiments employing a self-pulsing transmitter and a cathode-ray tube indicator. 4 refs. from 1918.

1120 Guerlac, Henry C., "The Radio Background of Radar." J.F.I., 250:285-308, Oct. 1950. 7 diags. A thorough survey of military needs, reflection of radio waves, short-wave experiments, the pulse-echo principle, ionospheric observations, ultrahigh frequency tubes, apparatus and techniques up to about 1931. 54 refs. from 1855, many multiple entries.

1121 Appleton, Edward V., and G. Builder, "Wireless Echoes of Short Delay." Proc. Phys. Soc., 44:76-78, Jan. 1, 1932. On the production of short radio-frequency pulses and applications in the investigation of wireless echoes.

1122 Watson-Watt, Robert A., and others, Applications of the Cathode-Ray Oscillograph in Radio Research. Contains much material related to the radio history of radar up to 1933. See 735.

1123 "The New Mystery Ray." Radio-Craft, 7:267, 299, Nov. 1935. 9 photos, 2 diags. Comments on recent reports of secret experiments with ultra-short waves for detecting aircraft being undertaken in Britain, Germany, Italy and the United States.

1124 Tuska, Clarence D., "Historical Notes on the Determination of Distance by Timed Radio Waves." J. F. I., 237:1-20; 83-102, Jan., Feb. 1944. 15 diags., mostly extracts from patents. A thorough survey from 1901 to about 1943. Radio investigations of the ionosphere, airplane transmission, studies of echo signals, recorders and indicators, pulse methods, altimeters. Bibliography, pp. 99-102, 81 entries from 1893.

1125 Tuska, Clarence D., "Pictorial Radio." J. F. I., 253:1-20; 95-124, Jan., Feb. 1952. 56 illus. On radar history, or graphic displays by means of reflected radio waves, with numerous diagrams of circuits and devices from patents up to 1950. 37 refs. from 1888.

B. Official and Group Histories of Radar

1126 "Radar Stories are Released by U. S. and Great Britain." Electronics, 16:274, 278, 280, 282, June 1943. First official news about radar. Essential text of the Joint Army-Navy and the British Information Service releases. Portrait of Sir Robert A. Watson-Watt.

1127 [U. S.]. "Navy's History of Radar." Electronics, 16:212, 214-219, July 1943. Navy press release, May 23, 1943. Portraits of Dr. A. H. Taylor and Mr. Leo C. Young. History from 1922 to 1941.

1128 Sanders, Frederick H., "Radar Development in Canada." Proc. IRE., 35:195-200, Feb. 1947. 12 photos. A sketch of early developments up to the end of 1940 with details of various radar equipments.

1129 Colton, Roger B., "Radar in the United States Army." Proc. IRE., 33:740-753, Nov. 1945. 36 photos and diags. History of the Signal Corps program from 1931, with description of developments at Fort Monmouth, N. J., and work done by participating companies up to 1941. Technical details are given with description of the SCR-268 and SCR-270 radars. No refs.

1130 Howeth, Linwood S., "Radar." History of Communications-Electronics in the United States Navy, Chap. 38, pp. 443-468. 9 photos. History up to 1941; early radar, principles, trials, service tests, development of naval radar, research by the Army Signal Corps, fire control radar and early wartime developments. 92 footnote refs., from 1922, to government letters, files, reports, personal statements and memoranda. See 305.

1131 [U.K.]. Radar, An Official History of the New Science. London, New York: British Information Service, June 1945. 30 pp. With technical descriptions and glossary of radar terms.

1132 [B.T.L.]. "Radar and the Bell System." Bell Lab. Rec., 23:325-328, Sept. 1945. Description of radar equipments developed by the Bell System.

1133 [M.I.T.]. "Longhairs and Short Waves." Fortune, 32:162-169, 206, 208, Nov. 1945. 11 photos. On the development of microwave radar at M.I.T. Radiation Laboratory.

1134 Brookfield, S.J., "Science in the Naval War." Discovery, 6(NS):363-373, Dec. 1945. Includes account of radar in the British navy. 13 illus. refer to radar and include equipment panels, ship antennas, magnetron, scope displays, proximity fused shell.

1135 [U.S.]. Radar, A Report on Science at War. By Joint Board on Scientific Information Policy for Office of Scientific Research and Development, War and Navy Departments. Washington: G.P.O., 1945. 53 pp.

1136 Vieweger, Arthur L., and Albert S. White, "Development of Radar SCR-270." Communications and Electronics Digest, pp. 19-23, Nov. 1959. 3 illus. and a table of characteristics.

1137 Kelly, Mervin J., "Radar and Bell Laboratories." Bell Tel. Mag., 24:221-255, Winter, 1945-1946. On wartime work at the Bell Laboratories with mention of AT&T, M.I.T., British work and some radar history.

1138 "Radar: A Story in Pictures." Bell Tel. Mag., 24: 257-282, Winter, 1945-1946. Pictorial description, including research, inventors, equipment, production and personnel.

1139 Lock, F.R., "Radar and Western Electric." Bell Tel. Mag., 24:283-294, Winter, 1945-1946. On production and manufacturing problems and operations.

1140 Tinus, William C., and William H. C. Higgins, "Early Fire-Control Radars for Naval Vessels." B. S. T. J., 25:1-47, Jan. 1946. 39 photos, diags. and graphs. Historical background, CXAS radar, Mark 1, 2, 3, 4 radars, with details of assemblies, installation and operation. 6 footnote refs.

1141 Berkner, Lloyd V., "Naval Airborne Radar." Proc. IRE., 34:671-706, Sept. 1946. 57 illus. Brief historical background, pp. 671-678. No refs.

1142 Stout, Wesley W., The Great Detective. Detroit: Chrysler Corporation, 1946. 99 pp., colored illus. Story of radar equipment and the company's activities during World War II. Popular historical text with numerous pictures of equipment and assembly.

1143 [U. K.]. "The Telecommunications Research Establishment, Malvern." Nature, 161:918, 919, June 12, 1948. A report on the first public exhibition of the work done at T. R. E., including microwave developments, radar techniques in meteorology, airborne radar, infrared and ultrasonics research.

1144 [M. I. T.]. Radiation Laboratory Series. Vol. 1, Radar System Engineering, 1947. Vol. 2, Radar Aids to Navigation, 1947. Vol. 3, Radar Beacons, 1947. Vol. 26, Radar Scanners and Radomes, 1948. Also other vols. on microwave equipment and systems. See 798.

1145 [U. S.]. Electronics Warfare--A Report on Radar Countermeasures. By Joint Board on Scientific Information Policy for Office of Scientific Research and Development, War and Navy Departments. Washington: G. P. O., 1945 (1951). 38 pp., illus.

1146 [U. K.]. "Origins of Radar. Background to the Awards of the Royal Commission." Wireless World, 58:95-99, March 1952. An account of radar development in Britain from 1935 through the war. Based on evidence presented to the Royal Commission on Awards to Inventors.

1147 Vieweger, Arthur L., "Radar in the Signal Corps." Trans. IRE., MIL-4:555-561, Oct. 1960.

C. General Accounts of Radar

1148 "Radio Location of Aircraft." Electrician, 126:356, June 20, 1941. Notes on official news release by Lord Beaverbrook.

1149 "Radiolocators." J. App. Phys., 12:511, July 1941. Editorial note on British Air Ministry news release and appeal for technical workers.

1150 Milbourne, S. C., "British Radio Combats Blitz, Radiolocator." Radio News, 26:6, 7, Aug. 1941.

1151 Page, Robert M., "The Early History of Radar." Proc. IRE., 50:1232-1236, May 1962. 9 illus. A quick glance at work in the United States from 1922 to 1941. 3 refs.

1152 Southworth, George C., "Waveguide Technique Goes to War." Forty Years of Radio Research, Chap. 11, pp. 179-192. On radar work from 1940 to 1942; lobe switches, phasing devices, and X- and K-band radar. See 926.

1153 Hightower, John M., "Story of Radar." U. S. Senate Documents. 78th Congress, 1st Session, 1943. 19 pp. An account of American developments reprinted from the daily press.

1154 Davis, B. G., "Who Invented Radar?" Radio News, 30:4, 70-72, July 1943. Editorial look at radar history.

1155 Cripps, Stafford, "The History of Radar." Engineer, 180:134, 135;153-155;170-172;190, 191, Aug. 17, 24, 31; Sept. 7, 1945. 6 illus.

1156 "Story of Radar." Electrical Times, 108:199-200, Aug. 16, 1945; Electrician, 135:265-266, Sept. 4, 1945. Early radar stations in Britain from 1935 through the war years are described. See also, "Achievements in Radiolocation," pp. 156-158, Aug. 17; and pp. 179, 180, 183-187, 198, Aug. 24.

1157 "Evolution of Radar." Engineering, 160:154, 155; 176, 177; 183-185; 205; 236, 237; 244, 245; 265; 285, Aug. 24, 31, Sept. 7, 14, 21, 28, Oct. 5, 12, 1945.

1158 "The B.T.H. Company and Radar." Engineer, 180: 212, 213, Sept. 14, 1945. 4 photos. Report on the contributions of the British Thomson-Houston Company in developing and manufacturing radar equipment from 1940. Also, Electrician, 135:265, 266, Sept. 14. 3 photos.

1159 Watson-Watt, Robert A., "Radar in War and Peace." Nature, 156:319-324, Sept. 15, 1945. General discussion with philosophical observations.

1160 Watson-Watt, Robert A., "Radar." Discovery, 6(NS): 281-290, Sept. 1945. 11 illus. Research, development and application of radar in Britain from 1935.

1161 Appleton, Edward V., "The Scientific Principles of Radiolocation." J. IEE., 92(Pt. 1):340-353, Sept. 1945. Includes evolution of radar from ionospheric observations, with reference to pioneer papers. 21 refs. from 1792.

1162 Cripps, Stafford, "The Pioneers of Radiolocation." Electronic Engineering, 17:680-686, Sept. 1945. Story of development with reference to some of the scientists.

1163 Dunsheath, Percy, "British Electrical Engineers and the Second World War." Engineer, 180:280, 281; 302-305, Oct. 12, 19, 1945. Summary of IEE Presidential address. Brief account of general technical activities including development of radar as well as communications.

1164 "Radar--The Technique." Fortune, 32:139-145, 196, 198, 200; "Radar--The Industry," 146, 200, 203, Oct. 1945. 10 illus. Popular description of physical principles and radar uses, with mention of U.S. agencies, contractors, specific types of equipment, and wartime episodes.

1165 DuBridge, Lee A., and Louis N. Ridenour, "Expanded Horizons." Tech. Rev., 48:23-26, 62, 64, 66, Nov. 1945. 6 photos. A general review of research at the M.I.T. Radiation Laboratory, of developments in radar components, techniques and equipment, and the expansion of microwave technology. See also 501, 507.

1166 "Radar Countermeasures." Engineering, 180:460, 461, Dec. 7, 1945. Various methods and techniques used to counter enemy radio and radar; with names, dates and events.

1167 Baxter, James P., "Radar and Loran." Scientists Against Time, Chap. 9, pp. 136-157. Also Chap. 10, pp. 158-169, Radar countermeasures. In addition to these chapters, radar is mentioned extensively throughout the book. See 507.

1168 Rowe, Albert P., One Story of Radar. Cambridge: University Press, 1948. 207 pp., plates. A personal record of work and activities on radio and radar at the Telecommunications Research Establishment in Britain during World War II.

1169 Page, Robert M., The Origin of Radar. New York: Doubleday, 1962. 196 pp., 8 plates, 40 illus. Radar chronology, pp. 183-187, 41 entries from 1832. A semitechnical description of basic electronics as well as radar. Devoted solely to American developments, with special reference to the Naval Research Laboratory. Up to the early 1940s.

1170 Watson-Watt, Robert A., Three Steps to Victory. London: Odhams Press, 1957; The Pulse of Radar, New York: Dial Press, 1959. History to 1945. See 462.

1171 Randall, John T., "Radar and the Magnetron." J. Roy. Soc. Arts, 94:303-312, discussion 312-323, April 12, 1946. Principles and development of radar and the cavity magnetron.

1172 Smith-Rose, Reginald L., "Radiolocation or Radar." RSGB Bull., 21:119-125, Feb. 1946. Covers historical developments from early ionosphere experiments up to centimeter radar.

1173 "The Historical Development of Radar." Proc. IRE Australia, 7:3-9, March 1946. An outline of radar development in England and the United States with mention of early radar in Germany and Japan.

1174 Llewellyn, Frederick B., "Radiolocation; Development of Radar." Engineer, 181:296, 297, March 29, 1946.

1175 Watson-Watt, Robert A., "The Evolution of Radiolocation." J. IEE., 93(Pt. 1):374-382, Sept. 1946. Lecture given at the Radiolocation Convention, London, March 26, 1946. A survey of British developments from

1935. Scientific background, technical details, the industry and milestones of evolution. Also, Discovery, 7(NS):113-115, April 1946; Engineering, 161:306-308, March 29, 1946; Engineer, 181:319-321; 330,331, April 5,12, 1946. See also 1176,1180.

1176 [I. E. E.]. "Proceedings of the Radiolocation Convention." J. IEE., 93(Pt. 3A):5-1620, Mar.-May, 1946. This important collection of papers describes radar developments in Britain during World War II. Also, "Bibliography of Radiolocation Convention Papers." J. IEE., 93(Pt. 1):478-480, Oct. 1946. 114 entries with brief descriptions. See also 1175,1180.

1177 DuBridge, Lee A., "The Story of Radar." Tech. Rev., 48:355-358,382, April 1946. 4 photos, 1 diag. A nontechnical account primarily about the activities at M.I.T. Radiation Laboratory.

1178 Selected Bibliography on Radar. Compiled by the staff of the Technological Institute Library. Evanston, Illinois: Northwestern University, June 1946. 41 pp., 547 entries. Historical background, 143 items from 1619, including U.S. patents. Some historical items in other sections; principles, terms, use in the war, industry, photography, scientific applications, scientists, transport, schools, prospects. Some entries are annotated.

1179 Wilkinson, R.I., "Short Survey of Japanese Radar." Elect. Eng., 65:370-377,455-463, Aug.-Sept., Oct. 1946. Report on Japanese radar, including history, development and military use.

1180 Llewellyn, Frederick B., "Greetings from the Institute of Radio Engineers." J. IEE., 93(Pt. 1):382-384, Sept. 1946. Survey of radar developments in an address at the Radiolocation Convention, 26 March, 1946. See also 1175,1176.

1181 Dunlap, Orrin E., Radar. What Radar Is and How It Works. New York: Harper and Brothers, 1946. 208 pp., 8 plates, 7 diags. Suggested reading, pp. 193-203, 145 books, articles and patents from 1893. For the general reader.

1182 Smith, Robert A., "Survey of the Development of Radar." J. IEE., 94(Pt. 1):172-178, April 1947.

11 illus., 28 refs.

1183 Fink, Donald G., Radar Engineering. New York: McGraw-Hill, 1947. 644 pp., 471 illus., 14 tables. Some chapter bibliographies, 63 items; also 56 footnotes, partly refs. Origins of radar, pp. 3-9, 8 footnotes. Bibliography, pp. 61, 62, 15 items.

1184 Smith, Robert A., "Radar." Electronics and Their Application in Industry and Research, Chap. 6, pp. 239-319. 55 illus. Technical treatment with a brief historical introduction. 50 refs. from 1928. See 799.

1185 Hornung, J. L., Radar Primer. New York: McGraw-Hill, 1948. 218 pp., 122 photos and diags. Bibliography, pp. 211-213, 39 books on radio, radar and television. Chap. 8, pp. 203-210, History of radar; brief survey of highlights.

1186 Wilkins, A. F., "The Story of Radar." Research, 6:434-440, Nov. 1953. Highlights of development in Britain from 1945.

1187 Garratt, Gerald R. M., "Birth of Radar." Electronic Engineering, 30:140-142, March 1958.

1188 Getting, Ivan A., "Radar." The Age of Electronics, Chap. 4, pp. 53-92. 30 photos and diags. Background of early developments in Britain and the United States and a review of progress from 1950. 2 refs. See 812.

1189 Skolnik, Merrill I., Introduction to Radar Systems. New York: McGraw-Hill, 1962. 648 pp., 381 illus., over 1000 chapter refs. History of radar development, pp. 8-14, with 17 refs. on p. 19.

13. RADIO

Early radio history is found in experiments and observations on electromagnetic waves; see Chap. 9, also 816 for a comprehensive survey of the early years. Contributions of some of the pioneers are contained in Sect. 4B, see especially 359, 369, 371. For the story of individual work, see Sect. 4C, especially the following: Armstrong, 379; de Forest, 396; Fessenden, 405; Hertz, 418; Marconi, 431; Maxwell, 436; Popov, 446. Other related items will be found in Sects. 3E, 3G.

A. General Histories

1190 "The History of Radio." Radio News, 5:1730-1732, 1816-1820, June 1924. 8 ports., one group photo, 3 other pictures. From 1827 to February 1924. Abstracted from the Radio Service Bulletin.

1191 Dowsett, H.M., Wireless Telephony and Broadcasting. London: Gresham Publishing Co., 1925, 2 vols. Vol. 1, 210 pp., 12 plates (ports., maps, charts), numerous other diags. and plates, some in color. App., pp. 203-210, licence and agreement, Post Office Telegraphs and the British Broadcasting Company, Jan. 18, 1923. Several historical chapters. Vol. 2, 233 pp., 19 plates (ports., apparatus), numerous other diags. and photos, some in color. Glossary, pp. 217-225. Biographical notes, pp. 227-230, 111 entries. Index for both vols., pp. 231-233. A few footnotes in both vols., otherwise not documented. A technical exposition and a broad survey of the radio art in Britain up to the mid-1920s.

1192 Morse, A.H., Radio: Beam and Broadcast. Its Story and Patents. London: Ernest Benn, 1925. 192 pp., 3 plates, 56 illus., about half from patents. First 100 pages is a survey up to the mid-1920s. App., pp. 101-186, contains abstracts from 21 important early patents. Bibliography, pp. 187, 188, has 30 entries. Numerous patent numbers and refs. in the text, others in footnotes. Also, Radio News, Vols. 6, 7, May-Sept. 1925. See 1369.

1193 Blake, George G., History of Radio Telegraphy and Telephony. London: Radio Press, 1926; Chapman and Hall, 1928. 425 pp., 246 diags. and photos. App., pp. 342-351, various items. Reference list, pp. 353-403, 1125 entries, mostly papers, articles and patents. More than 150 footnotes and many other names and dates in the text. Early history of electrical signalling and the telephone, early wireless communication systems, electromagnetic wave experiments, coherers, magnetic detectors, electrolytic and crystal detectors, early radio telegraphy and telephony, spark, arc, and other high-frequency generators, microphones, facsimile, remote control, thermionic valves and circuits, direction finding. "The book is intended to provide a useful work of reference wherein will be found many now almost forgotten 'wireless' schemes and inventions" and is, indeed, a collection of entries rather than a formally organized history. The wide variety of miscellaneous materials and the extensive references relate mainly to components, circuits, experiments, proposals and systems.

1194 Archer, Gleason L., History of Radio to 1926. New York: The American Historical Society, 1938. 421 pp., 20 plates. App., pp. 375-399, contains various agreements and a list of broadcasting stations, 1920-1922, pp. 393-397. Detailed index, pp. 401-421. Treats early developments, in the beginning and during World War 1; origin of RCA and contemporary companies, early broadcasting and personalities, litigation and rivalries, early network activities, wavelength problems, Federal matters. Heavy emphasis on the American scene. 175 footnotes, mostly refs. See also 1417.

1195 One Hundred Years of Radio Development. Compiled from U. S. Government Sources. Washington Bureau, n. d. 4 pp. Brief entries from 1827 to late 1927.

1196 Clarkson, R. P., The Hysterical Background of Radio. New York: J. H. Sears, 1927. 257 pp. A nontechnical and light survey of electrical progress from the earliest times. Inventors, experimenters and organized research are discussed.

1197 Dunlap, Orrin E., The Story of Radio. New York: Dial Press, 1927. 226 pp., 15 plates. A popular account for the general reader. No refs.

1198 "Important Events in Radio--Peaks in the Waves of Wireless Progress, 1827 to 1928." Radio Service Bulletin, 141:16-26, Dec. 1928. Washington: Department of Commerce, Radio Division.

1199 Harbord, James G., "Radio." A Century of Industrial Progress, pp. 552-564. A brief historical view including marine communications, broadcasting and facsimile. See 336.

1200 Schubert, Paul, The Electric Word: The Rise of Radio. New York: Macmillan, 1928. 311 pp. A readable narrative of radio developments that lightly touches scientific matters as well as politics and business. No index, no refs.

1201 Codel, Martin (ed.), Radio and Its Future. New York: Harper and Brothers, 1930. 349 pp., 8 plates. Pt. 1, pp. 3-91, Broadcasting; Pt. 2, pp. 95-182, Communications; Pt. 3, pp. 185-215, Industry; Pt. 4, pp. 219-264, Regulation; Pt. 5, pp. 267-327, Some scientific and other considerations. A broad review for the layman with a little historical background.

1202 Marconi, Guglielmo, "Thirty-Seven Years of Radio Progress." Radio News, 13:(552), 553, 554, 608, 610, Jan. 1932. 6 illus. Talk by Marconi during the Faraday Centennial in London that was transmitted to the United States.

1203 Taylor, J. E., "Notes on Wireless History." J. POEE., 25:295-299, Jan. 1933. A brief survey of radio progress.

1204 Eccles, William H., Wireless. London: Thornton Butterworth, 1933. 256 pp., 82 illus. Bibliography, pp. 245-249, 59 items. A semitechnical and popular survey from the experiments of Hertz to the "progress of recent years." This useful book by a recognized authority includes various pieces of historical interest and a few patent refs.

1205 Vyvyan, Richard N., Wireless Over Thirty Years. London: George Routledge, 1933. 256 pp., 16 plates, 12 diags. App., pp. 239-256, 12 diags. of Marconi's early circuits, 2 lists of stations, 6 tables of frequency bands, stations, services, allocations, etc. A general and nontechnical account with personal reminiscences by a close associate of Marconi. Early history, transatlantic wireless, high-power stations, histories of Imperial Wireless and beam development, wireless during World War 1, commercial development, broadcasting, and the work of the British Post Office. No index, no refs.

1206 O'Dea, William T., Radio Communication: Its History and Development. London: H. M. S. O., 1934. 95 pp., 34 plates. Refs., p. 92, 14 books, 4 research reports and 15 other publications. An authoritative survey without specific refs. Objects in the Science Museum collection are noted in the text.

1207 Espenschied, Lloyd, "The Origin and Development of Radiotelephony." Proc. IRE., 25:1101-1123, Sept. 1937. 8 illus. Bibliography, pp. 1121-1123, 45 items from 1880. A unified account primarily devoted to American contributions; 1912 through World War 1, post-war experiments, early broadcasting, radiotelephony of the 1920s, and ultrahigh frequencies.

1208 Reck, Franklin M., Radio from Start to Finish. New York: Thomas Y. Crowell, 1942. 160 pp. Diags. and ports. Radio history from 1895 to the early 1920s is surveyed in pp. 1-65. The remainder treats station operations and networks. Television, FM and military radio are mentioned. A beginner's book.

1209 Kelsey, Elizabeth (comp.), Trail Blazers to Radionics and Reference Guide to Ultra High Frequencies. Chicago: Zenith Radio Corp., 1943. 29, 56 pp. Pt. 1, 45 brief biographies, from Pythagoras to Farnsworth, with refs. Pt. 2, bibliography, 726 articles and books, in 21 categories.

1210 Crawley, Chetwode, G., "Fifty Years of Radio-Communication." Discovery, 7(NS):150-152, May 1946. 2 views, 5 ports.

1211 McNicol, Donald, Radio's Conquest of Space. The Experimental Rise in Radio Communication. New York: Murray Hill Books, 1946. 374 pp., 69 illus., including numerous ports. A semitechnical record of radio progress from the mid-19th century, more or less in chronological order. Although some emphasis is placed on American work, this is one of the best surveys of inventive achievement that is both readable and reliable. 129 footnotes, mainly refs. to papers and patents, other refs. in the text.

1212 Maclaurin, William R. (with R. Joyce Harman), Invention and Innovation in the Radio Industry. New York: Macmillan, 1949. 304 pp., 16 plates, 22 diags., 38 tables. App. 1, pp. 266-268, Elements of modern radio communication. App. 2, pp. 269-291, Radio patent litigation. Bibliography, pp. 292-298, lists 145 books and articles; biography, economics and business, patents, radio, television, science and industrial research. A study of struggles, connivances, litigation, progress and failure of and between independent inventors and industrial organizations. Thorough treatment, with supporting data and much critical analysis, of the process and nature of radio invention and innovations set in the mosaic of contemporary economic, social, financial and personal circumstances. Includes television history. Thoroughly documented with more than 500 footnotes.

1213 de Forest, Lee, "A Half Century." Proc. IRE., 38: 1379, Dec. 1950. Radio events of 50 years are skimmed in this guest editorial.

1214 Read, Oliver, "From Catswhiskers to Color." Radio and Television News, 52:31, 105-107, July 1954. Brief historical comments.

1215 Kennedy, T. R., "From Coherer to Spacistor." Radio-Electronics, 29:44-59, April 1958. 24 illus. and list of Gernsback Electronic Publications 1908-1958. A 50-year history of Gernsback's radio and allied publications with highlights of events in radio progress.

1216 Rodunskaya, I., and M. Zhabotinsky (E. Felgenhauer, trans.), Radio Today. Moscow: Foreign Languages Publishing House, 1959. 263 pp., 46 illus. The Russian story of the development of electronics. Television, radio, radar, radio astronomy, radio spectroscopy, computers, radio in industry, semiconductor devices and radio in the conquest of space. A popular text without documentation, although many names, dates and inventions are mentioned. Scant reference to work done outside the U. S. S. R. No index.

1217 The Old Timer's Bulletin. Holcomb, N. Y.: Antique

Wireless Association, 1960. Vol. 10, No. 4, March 1970. Index to Vols. 1-10, classified by issue sequence under 17 topical headings.

1218 Buff, Christopher, "Radio Receivers--Past and Present." Proc. IRE., 50:884-891, May 1962. 9 illus., 10 footnote refs., 4 items in a bibliography. Highlights of important techniques, circuits and developments over 50 years.

1219 Weldon, James O., "Transmitters." Proc. IRE., 50:901-906, May 1962. 6 illus. A brief survey from 1901.

1220 Pratt, Haraden, "Sixty Years of Wireless and Radio." IEEE Spectrum, 6:41-47, Nov. 1969. 6 photos. Autobiographical reminiscences selected from personal memoirs.

1221 Sivowitch, Elliot N., "A Technological Survey of Broadcasting's "Pre-History," 1876-1920." J. Broadcasting, 15:1-20, Winter 1970-1971. Highlights of experimental work from the invention of the telephone to the advent of commercial broadcasting. Wired broadcasting, aerial conduction, the work of Stubblefield, Edison, Dolbear, E. Thomson, Stone, Fessenden, de Forest, West Coast experiments, Sarnoff and RCA. 37 refs.

B. Wireless Telegraphy to 1900

Many attempts to conduct electricity without wires for signaling purposes, made during the 19th century, are described in Fahie's History, 1234. Apart from induction methods and experiments by Loomis (1222) on atmospheric conduction, other experiments by Houston, Edison, and Dolbear were concerned with electromagnetic-wave radiation, although not recognized as such until the end of the century. Details of these and similar experiments are related in several items listed in Sect. 9A, especially 866-871, 874, 878, 879. See also 369, 816.

1222 Winters, S. R., "The Story of Mahlon Loomis, Pioneer of Radio." Radio News, 4:836, 837, 966-978, 980, 981, Nov. 1922. 11 illus., including text of original patent and copies of drawings. Extracts from articles and records, with refs. in the text, from 1868. See also 429.

1223 "Weak Sparks." Sci. Am., 33:385, Dec. 18, 1875. Editorial comment on Edison's experiments of November 22 on spark effects which soon became known as an "etheric force." A series of about 20 articles, letters and comments ensued. See next two items, also 866-870.

1224 "The New Phase of Electric Force." Sci. Am., 33: 401, Dec. 25, 1875. 3 illus., including view of experimental apparatus. (Also editorial note, p. 400.)

1225 Houston, Edwin J., and Elihu Thomson, "Experiments With Ruhmkorff Coil and Spark." J. F. I., 101:270-274, April 1876; Sci. Am. Supp., 1:326, May 20, 1876. Also, "Notes on the Phenomena of Induction--Etheric Force." English Mechanic, 22:549, 550, Feb. 11, 1876; and "The So-called Etheric Force," 23:242, 243, May 19, 1876. For earlier experiments, see J. F. I., 91:417-419, 1871.

1226 [Hughes, David E.]. Fahie, John J., "Researches of Prof. D. E. Hughes, F. R. S., in Electric Waves and their Application to Wireless Telegraphy, 1879-1886." A History of Wireless Telegraphy, App. D, pp. 289-296. Letters by Crookes, Fahie and Hughes on the latter's experiments of 1879, which were not made public for 20 years. See 1234, also Electrician, 43:40, 41, May 5, 1899; Elect. Review, 44:883-885, 1899; and 874 (3rd ed., pp. 88-94).

1227 Dolbear, Amos E., "On the Development of a New Telephonic System." J. Soc. Telegraph Engineers and Electricians, 11:129-149, 1882.

1228 Branly, Eduoard, "Variations of Conductivity Under Electrical Influence." Electrician, 27:221-223; 448, 449, June 26, Aug. 21, 1891. See also 874, 3rd ed., pp. 95-108; 1234, pp. 276-288. A pioneer paper on the coherer. 2 illus.

1229 Crookes, William, "Some Possibilities of Electricity." Fortnightly Review, 51:173-181, Feb. 1892. This famous paper contains a prediction of the means for radio telegraphy: transmitter, receiver, directive antenna, circuit tuning and selectivity.

1230 Lodge, Oliver J., "The History of the Coherer Principle." Electrician, 40:87-91, Nov. 12, 1897. See also 874, 3rd ed., pp. 73-87.

1231 Marconi, Guglielmo, "Wireless Telegraphy." Proc. Roy. Inst., 16:247-256, 1899; J. IEE., 28:273-297, 300-316, 1899. A survey of his work up to the beginning of 1899.

1232 Kerr, Richard, Wireless Telegraphy Popularly Explained. London: Seeley, 1898. 5th ed., 1901. 116 pp., 17 illus., ports., and diags. Preface by William H. Preece. One of the first "popular" books, with background on science, vibrations, induction experiments, and Hertzian waves. No refs., no index.

1233 Thompson, Silvanus P., "Telegraphy Across Space." J. Roy. Soc. Arts, 46:453-460, 1898. Also, Smithsonian Institution, Annual Report for 1898, pp. 235-247 (Washington, 1899). A brief sketch of communication without wires by the conduction, induction, and electric wave methods.

1234 Fahie, John J., A History of Wireless Telegraphy 1838-1899. Edinburgh, London: Blackwood; New York: Dodd, Mead, 1899. 325 pp., frontispiece and 65 diags. Most of the book is devoted to "bare wire proposals for subaqueous telegraphs," as the subtitle states. G. Marconi's Method, pp. 177-245. App. E, pp. 296-320, Reprint of Signor G. Marconi's Patent (see 1334). D. E. Hughes' researches, pp. 289-296 (see 1226). 2nd ed., 1901, 348 pp. See also 1528.

1235 Pickard, George W., "History and Present Development of Wireless Telegraphy." Sci. Am. Supp., 49:20096-20098, Jan. 13, 1900.

1236 Bottone, Stanley R., Wireless Telegraphy and Hertzian Waves. London: Whittaker, 1900. 116 pp., 35 illus., mainly diags. A simple account with some constructional details for the amateur. Historical matter, pp. 13-59. A pioneer practical book in this field.

1237 Howe, G. W. O., "Alexander S. Popov." Wireless Engineer, 25:1-5, Jan. 1948. A review of Popov's life and work with observations on the invention of radio communication and the rival claims of Popov and Marconi. Also 25:135-137, May 1948; 26:141, 142; 249, 250, April, Aug. 1949.

1238 Süsskind, Charles, "Popov and the Beginnings of Radiotelegraphy." Proc. IRE., 50:2036-2047, Oct. 1962. 4 illus. Critical review of the Marconi-Popov controversy. Thoroughly documented, 83 refs. See also, Proc. IEEE., 51:473, 474, March 1963; 53:162-164, Feb. 1965. Also, San Francisco: San Francisco Press, 1962. 30 pp., 4 illus., 83 refs. New ed., 1970.

C. Wireless Telegraphy and Telephony, 1901 to 1920

Highlights of this period include the first transatlantic transmission by Marconi in 1901, development of high-power equipment, inception of practical electron-tube technology, growth of radiotelephony, and the world-wide application of arc, spark, and high-frequency alternators for long-distance commercial radiotelegraph circuits by the end of World War 1. Other items relevant to this period will be found in Sects. 10A, 10B, especially 949, 955, 970, 971, 975, 977, 979. See also Sect. 13E for related entries concerning circuits and patents.

1239 Marconi, Guglielmo, "Signals Across the Atlantic." Electrical World, 38:1023-1025, Dec. 21, 1901. Also, Electrician, 48:329 ("Notes"), Dec. 20, 1901.

1240 "Transatlantic Wireless; Thirtieth Anniversary of First Communication." Electrician, 107:847, Dec. 18, 1931.

1241 Smith-Rose, Reginald L., "Anniversary of Transatlantic Radio." Nature, 168:980, Dec. 8, 1951. Also, Electrical Times, 120:1070, 1071, Dec. 13, 1951; Engineer, 192:772, 773, Dec. 14, 1951; Electrician, 147:1965, 1966, Dec. 21, 1951. 5 photos.

1242 Coulson, Thomas, "Radio's Memorable Anniversary." J. F. I., 253:287-292, April 1952. On the 50th anniversary of transatlantic radio. A talk presented at WGY, Schenectady, Dec. 12, 1951.

1243 Marconi, Guglielmo, "The Progress of Electric Space Telegraphy." Proc. Roy. Inst., 17:195-210, 1902. Also, Electrician, 49:388-392, June 27, 1902. 3 circuit diags. Survey of his work from 1900.

1244 Maver, William, "Wireless Telegraphy--Its Past and Present Status and Its Prospects." Smithsonian Institution, Annual Report for 1902, pp. 261-274 (Washington, 1903). 13 illus., including 4 plates. A summary of events from the early 1890s. Also, Cassier's Mag., 21:213-227, Jan. 1902.

1245 Sewall, Charles H., Wireless Telegraphy: Its Origin, Development, Inventions, and Apparatus. New York: D. Van Nostrand; London: Crosby, Lockwood, 1903. 229 pp., 85 illus. An overview of history, principles, theory, apparatus and practical operation. This was the first book on wireless telegraphy published in the United States. 4 Apps., patents of Dolbear, Lodge, Marconi, Tesla.

1246 Maver, William, "Progress in Wireless Telegraphy." Smithsonian Institution, Annual Report for 1903, pp. 275-280 (Washington, 1904). 3 diags. On recent work, particularly with coherers and other detectors. Also, Cassier's Mag., 24:165-168, June 1903.

1247 Fleming, John A., "Hertzian Wave Wireless Telegraphy." Pop. Sci. Mon., 63:97-122, 193-208, 362-372, 439-452, 551-562; 64:53-65, 152-164, June 1903-Jan. 1904. 26 illus. A less technical version of the Cantor Lectures at the Society of Arts, March 1903. 77 footnotes. Later incorporated in Fleming's Principles (1253).

1248 Poincaré, Henri, and Frederick K. Vreeland, Maxwell's Theory and Wireless Telegraphy. New York: McGraw; London: Constable, 1904. Pt. 1, Maxwell's Theory and Hertzian Oscillations by Poincaré, trans. by Vreeland. Pt. 2, The Principles of Wireless Telegraphy, by Vreeland. 255 pp., 145 diags. Contemporary techniques and theories set against a historical background. 77 footnotes, mainly refs.

1249 Story, Alfred T., The Story of Wireless Telegraphy. London: Newnes; New York: D. Appleton, 1904. 215 pp., 57 illus. Conduction and induction experiments, pp. 18-100. Experiments of Hertz, Lodge and their contemporaries, pp. 101-129. Marconi's work and other pioneer systems, pp. 130-211. No refs.

1250 Marconi, Guglielmo, "Recent Advances in Wireless Telegraphy." Smithsonian Institution, Annual Report for 1905, pp. 131-145 (Washington, 1906). 13 illus. From

Royal Institution paper, March 3, 1905.

1251 Collins, A. Frederick, Wireless Telegraphy. Its History, Theory and Practice. New York: McGraw, 1905. 299 pp., 332 illus. A brief historical introduction is given with 13 of 20 chapters. A thorough coverage of electrical theory and contemporary practice with chapters on oscillators, induction coils, interruptors, detectors, transmitters, receptors, subsidiary apparatus, aerials and earths, syntonization, and wireless telephony. 139 footnotes.

1252 Collins, A. Frederick, Manual of Wireless Telegraphy and Telephony. New York: John Wiley; London: Chapman and Hall, 1906, 1909. 3rd ed., 1913. 300 pp., 129 plates, photos and diags., folding chart of North Atlantic Communications. Theory, commercial apparatus, transmitters, receivers, wiring diagrams, adjustments and operations, different systems, antennas and components. Chap. 12, pp. 244-259, Wireless telephony, by Newton Harrison. Book list, pp. 261-270, 41 items, including foreign-language books and government publications, with brief descriptions.

1253 Fleming, John A., The Principles of Electric Wave Telegraphy. London: Longmans, Green, 1906. 671 pp., 302 illus., 7 folding plates. App. 1, pp. 635-637, Wireless telegraphy act, 1904. App. II, pp. 638-641, Bibliography; 32 books, 44 original papers and lectures, from 1893. App. III, pp. 642-657, British patents relating to improvements in electric wave wireless telegraphy granted between 1896 and 1906, 273 items listed by year with name, number, date and title. Chap. 7, pp. 419-463, The inception of electric wave telegraphy, is a detailed historical record from early ideas up to 1905, with emphasis upon Marconi's progress and the work of Lodge and Fleming. Chap. 8, pp. 465-544, The development of electric wave telegraphy, deals thoroughly with progress by all workers in the field from 1897. Work in the United States is treated in detail with descriptions of some 170 patents of Fessenden, de Forest, Stone, Shoemaker. European contributions by Slaby and von Arco, Braun and the Telefunken company are similarly described along with patent references and illustrations of the various ideas and apparatus. 355 footnotes, mainly refs., some multiple entries. A foremost treatise and valuable record. 2nd ed., 1910. 906 pp., 478 illus. 4 Apps., including bibliography, pp. 890-892, 45 entries. Chap. 7, pp. 511-635, The evolution of electric wave telegraphy. Chap. 10, pp. 844-865, Radiotelephony. 3rd ed.,

1916; 4th ed., 1919.

1254 Fessenden, Reginald A., "Recent Progress in Wireless Telephony." Sci. Am., 96:68, 69, Jan. 19, 1907.

1255 Erskine-Murray, James, A Handbook of Wireless Telegraphy. London: Crosby, Lockwood; New York: D. Van Nostrand, 1907. 322 pp., plates, diags. and tables. Theory, practice and useful data for engineers, operators and students. Covers early history with details of apparatus and trials. Later eds.

1256 Fessenden, Reginald A., "Wireless Telephony." Proc. AIEE., 27:558-627, July 1908. Also Smithsonian Institution, Annual Report for 1908, pp. 161-195 (Washington, 1909). 39 illus. (20 plates). Brief history of wireless signaling and wireless telephony, description of methods, apparatus, circuits and operation, discussion of possibilities. 98 footnote refs. from 1847, including many patents. An important early paper.

1257 Fleming, John A., An Elementary Manual of Radiotelegraphy and Radiotelephony for Students and Operators. London, New York: Longmans, Green, 1908. 340 pp., numerous diags. Includes an account of early history.

1258 Ruhmer, Ernest (James Erskine-Murray, trans.), Wireless Telephony in Theory and Practice. London: Crosby, Lockwood; New York: D. Van Nostrand, 1908. 237 pp., numerous diags. Bibliography, pp. 215-217.

1259 Marconi, Guglielmo, "Wireless Telegraphic Communication." Nobel lecture, Dec. 11, 1909. See 213, pp. 196-222. Biography, pp. 223-225. Presentation speech, pp. 193-195. 25 photos, maps, diags. 22 refs. from 1896. (With K. F. Braun, see 885.)

1260 Pierce, George W., Principles of Wireless Telegraphy. New York: McGraw-Hill, 1910. 350 pp., 235 illus., 15 tables. Technical exposition with much historical material. Pages 140-214, on detectors. Wireless telephony, pp. 305-311. 131 footnotes, mainly refs.

1261 Fleming, John A., "Wireless Telegraphy." Encyclopaedia Britannica, 1911. 11th ed., 26:529-541. 16 diags. Early conduction and induction methods, space telegraphy, Marconi's system, transmitters and antennas, detectors and receivers, transatlantic telegraphy, directive telegraphy, instruments and measurements. 49 footnotes, list of refs.

1262 Taylor, J. E., "Early History of Wireless Telegraphy." J. POEE., 2:1-6, 81-87, 157-162, 291-299; 3: 47-51, 124-127, 260-262, 347-351; 4:20-24. 1909-1912.

1263 [Marconi Company]. The Year Book of Wireless Telegraphy and Telephony. London: Wireless Press. First published in 1913 by The Marconi Press Agency. Regular features include: Calendar, historical record, conventions, world list of stations, formulas, glossary, dictionary of technical terms, tables of useful data, patents, companies, biographical notices, obituary, list of books and journals grouped by countries, directory of amateur societies, call letters, code signals, laws and regulations, etc. About 100 pages contain several articles by noted writers. Illustrations (usually about 20 plates) consist of portraits and views of installations and equipment. Advertisements for services, equipment and supplies usually occupy about 100 pages. Each issue (1000-1500 pp.) is a mine of contemporary information, despite the predominance of Marconi Company interests.

1264 de Forest, Lee, "Recent Developments in the Work of the Federal Telegraph Company." Proc. IRE., 1: 37-57, Jan. 1913; Proc. IEEE., 51:426-433, March 1963. 23 photos and diags. Description of the Company installations and operations, Poulsen arc generators, and other transmitting equipment. Service performance, frequency-shift keying and selective fading are also discussed.

1265 Fleming, John A., The Wonders of Wireless Telegraphy. London: Society for Promoting Christian Knowledge, 1913. 280 pp., 55 illus. This explanation for the nontechnical reader deals with history and includes some early circuits.

1266 Hogan, John V. L., "A New Marconi Transatlantic Service." Electrical World, 64:425-428, Aug. 29, 1914. 7 photos, 3 refs. Brief historical background and description of the Marconi station at Carnarvon, Wales.

1267 Hogan, John V. L., "The Goldschmidt Transatlantic Radio Station." Electrical World, 64:853-855, Oct. 31, 1914. 4 photos, 2 diags., 2 refs. Description of the

Goldschmidt 100-Kw high-frequency alternator and the plant at Tuckerton, New Jersey.

1268 Gibson, Charles R., Wireless Telegraphy and Telephony Without Wires. London: Seeley, Service; Philadelphia: J. B. Lippincott, 1914. 156 pp., 9 plates, 19 diags. A popular history. Chap. 15, pp. 141-148, The evolution and development of wireless telegraphy, is a chronological record from 1831 to 1913, 41 entries.

1269 "Wireless Telephony. New Short Distance Apparatus." Wireless World, 2:589-591, Dec. 1914. Photograph of Marconi tube set, a combined transmitter and receiver. Report on Marconi's experiments in Italy earlier that year. A brief announcement appeared earlier in Wireless World, 2:35, April 1914.

1270 Fleming, John A., "Radiotelegraphy: A Retrospect of Twenty Years." Electrician, 77:831-836, Sept. 15, 1916. 4 photos, 2 diags.

1271 Round, Henry J., "Wireless Telephony." The Year Book of Wireless Telegraphy and Telephony (1915), pp. 572-582. 3 plates, 4 diags. An early description of the reaction valve (triode oscillator) and its use in wireless telephony. Experiments of 1914. See 1263.

1272 Zenneck, Jonathan (A. E. Seelig, trans.), Wireless Telegraphy. New York: McGraw-Hill, 1915. 1st German ed., 1909, 2nd ed., 1913. 443 pp., 461 illus. 13 tables, pp. 384-407. Bibliography and notes on theory, pp. 408-428, 355 entries referenced to the text and to some of the 370 footnotes, most of which supplement the text. A first-rate engineering text with a good deal of historical material.

1273 Coursey, Philip R., "The Methods Employed for the Wireless Communication of Speech." Wireless World, 4:47-52, 90-96, 216-223, 305-313, 385-387, April-Aug. 1916. 20 illus. Technical paper with historical treatment; arc, spark, vacuum tube and alternator systems, modulation, receivers. 29 refs.

1274 Bucher, Elmer E., Practical Wireless Telegraphy. New York: Wireless Press, 1917. Revised ed., 1921. 336 pp., 334 photos, diags., charts. A practical textbook for "students of radio communication." Useful historical material. Not documented.

1275 Lansford, Willis R., "Aircraft Communication in World War I." Radio News, 29:50-54, 208-216, June 1943. 10 photos of early equipment. Historical survey from 1910 to October 1917.

1276 Eccles, William H., Wireless Telegraphy and Telephony. A Handbook of Formulae, Data and Information. London: Benn Brothers; New York: D. Van Nostrand, 2nd ed., 1918. 514 pp., 434 illus., 2 folding charts. Tables, pp. 1-52. Formulae, pp. 53-115. Antennas and waves, pp. 116-185. Generators and plant, pp. 186-269. Detectors and receivers, pp. 270-357. High-frequency circuits, pp. 358-383. Systems and stations, pp. 384-484. Glossary, pp. 487-514. There is much historical matter in the text, with names, dates and other particulars, especially patents.

1277 Goldsmith, Alfred N., Radio Telephony. New York: Wireless Press, 1917. 247 pp., 226 photos, diags. and charts. The first comprehensive textbook of the period with numerous circuit diagrams and pictures of apparatus. Generators, pp. 21-126, includes arc, spark, vacuum tube and alternator systems. Modulation, pp. 127-204, includes microphones, vacuum tubes, magnetic amplifiers. Antennas and reception, pp. 205-229. 13 footnotes, other refs. in the text.

1278 Burrows, Arthur R., "Wireless Possibilities." The Year Book of Wireless Telegraphy and Telephony (1918), pp. 952-962. Brief history from Marconi's early work with speculations concerning future applications and services, especially the wireless distribution of public information and music, audible advertisements and other aspects of broadcasting. See 1263.

1279 Craft, E. B., and Edwin H. Colpitts, "Radio Telephony." Trans. AIEE., 38:305-343, Feb. 1919. 43 photos, graphs, diags. Pt. 1, Historical; Pt. 2, Commercial development. Largely about AT&T, Western Electric, and army and navy apparatus through World War I. An engineering approach.

1280 Eccles, William H., "Wireless Telegraphy and Telephony during the War." J. IEE., 59:77-84, Dec. 1920. Illus. Also, Engineering, 110:783-785, Dec. 10, 1920.

1281 Cusins, A. G. T., "Development of Army Wireless during the War." J. IEE., 59:763-770, July 1921.

1282 Eccles, William H., The Influence of Physical Research on the Development of Wireless. London: Institute of Physics, 1930, 18 pp. Also "Physics in Relation to Wireless." Nature, 125:894-897, June 14, 1930. Brief survey of highlights to 1918. No refs.

1283 Coursey, Philip R., Telephony Without Wires. London: Wireless Press, 1919. 414 pp., 249 illus. Brief list of achievements in wireless telephony, pp. 365, 366, 28 entries from 1885. List of refs., pp. 367-396, has 695 entries of articles and patents arranged by chapters. Name index, pp. 397-401, refers to the reference list as well as text pages. A comprehensive coverage of techniques, devices and developments with emphasis on the period 1905 to 1918.

1284 The Principles Underlying Radio Communication. Signal Corps, U. S. Army. Washington: G. P. O., 1919. 355 pp., illus. 2nd ed., 1921, 619 pp., 294 illus. Elementary electricity, dynamo-electric machinery, radio circuits, electromagnetic waves, transmitting and receiving apparatus (exclusive of electron tubes), electron tubes in radio communication. App. 8, pp. 568-577, lists 27 radio books, 33 Bureau of Standards radio publications, and 33 pamphlets on communications issued by the Signal Corps. This Radio Communication Pamphlet No. 40 was prepared by the Bureau of Standards for the Signal Corps. There are various footnotes and many references in the text. Useful historical material.

1285 Elwell, Charles F., "The Poulsen System of Radiotelegraphy: History of Development of Arc Methods." Electrician, 84:596-599, May 28, 1920. 7 illus. Survey from 1902.

1286 Marriott, Robert H. (ed. Haraden Pratt), "Radio Ancestors--an Autobiography." IEEE Spectrum, 6:52-61, June 1968. 4 illus. A condensation of miscellaneous matters and personal memories from the early days up to 1919. 47 refs. from 1870.

D. Radio from 1921

The 1920s witnessed the rise of radio as a major industry centered on electron-tube technology (Chap. 10) and the world-wide spread of broadcasting (Chap. 6). Popular books and articles, surveys and reminiscences and works on amateur radio are included in this section. A few engineering papers and textbooks have been selected (from a mass of similar material) for their historical treatment or portrayal of contemporary equipment and practices. See also Sect. 1E, 52-62. For related material, see Chap. 3; 268, 283-285, 287, 295, 336, 337; Chap. 6; 609, 615, also 1012. See also Sect. 13F for related entries concerning circuits and patents. A useful survey of the radio art at the beginning of the decade is in The Electric Journal, April 1921. Three of the 18 articles are given below.

1287 Alexanderson, Ernst F. W., "Central Stations for Radio Communication." Proc. IRE., 9:83-94, Feb. 1921. 6 photos and diags. Brief historical survey, discussion of high-frequency alternators, and description of the Radio Central Station and its antenna system.

1288 Chubb, L. W., and C. T. Allcutt, "The Foundations of Modern Radio." Electric Journal, 18:120-122, April 1921. Brief survey of four notable landmarks; continuous wave transmission, heterodyne reception, the vacuum tube, the feedback circuit.

1289 Little, D. G., "Continuous Wave Radio Communication." Electric Journal, 18:124-129, April 1921. 6 photos of equipment. Arc generators, high frequency alternators, vacuum tubes, and wartime developments.

1290 Brackett, Q. A., "Radio Arc Transmitters." Electric Journal, 18:142-146, April 1921. 5 photos of equipment, 1 diag. History and technical description.

1291 Morecroft, John H., Principles of Radio Communication. New York: John Wiley; London: Chapman and Hall, 1921. 935 pp. 788 photos, diags. and graphs. There are chapters on spark telegraphy, vacuum tubes and circuits, continuous-wave telegraphy, radio telephony, antennas and radiation, wavemeters, amplifiers, and experiments, in addition to others on basic electrical principles at high frequencies. A comprehensive treatise noteworthy for the many graphs and photos of waveforms obtained by the mirror oscillograph. No refs. except occasional footnotes. 2nd ed., 1927, 1001 pp. 3rd ed., 1933, 1084 pp. A basic text.

1292 Hooper, Stanford C., "Keeping the Stars and Stripes in the Ether." Radio Broadcast, 1:127-132, June 1922. Background of the Marconi Companies (British, U.S and the negotiations that led to the formation of Radio Corporation of America.

1293 Morecroft, John H., "What Everyone Should Know About Radio History." Radio Broadcast, July, Aug. 1922. Illus.

1294 Yates, Raymond F., and Louis G. Pacent, The Complete Radio Book. New York: P.F. Collier, 1922. (Vol. 17 of the Popular Science Library.) 330 pp., 77 illus., including 17 plates and one folding map. Popular, slightly historical. Chap. 17, pp. 286-315, lists 47 noteworthy persons with brief biographical details. No refs.

1295 Blake, George G., "Historical Notes on Radio Telegraphy and Telephony." Wireless World, 12:253-256, 286-293, 1922.

1296 Lodge, Oliver J., "The Origin or Basis of Wireless Communication." Nature, 111:328-332, March 10, 1923.

1297 Burghard, George E., "Eighteen Years of Amateur Radio." Radio Broadcast, 3:290-298, Aug. 1923. 7 illus. Reminiscences from 1905. Talk presented to the Radio Club of America by the club president.

1298 Elwell, Charles F., The Poulsen Arc Generator. London: Ernest Benn; New York: D. Van Nostrand, 1923. 192 pp., 150 illus. Bibliography, pp. 177-190, 127 entries. Chap. 2, pp. 22-26, History.

1299 Hogan, John V. L., The Outline of Radio. Boston: Little, Brown, 1923. New ed., 1925. 268 pp., 16 plates, 68 illus. Glossary, pp. 233-249. Bibliography, pp. 251-257, gives 10 books with comments, and 7 other items. Chap. 1, pp. 3-25, is a brief historical review. No refs.

1300 de Forest, Lee, "Quarter Century of Radio." Electrical World, 84:579, 580, Sept. 20, 1924.

1301 Elwell, Charles F., "Radio: Its Past, Present, and Future." J. Roy. Soc. Arts, 74:757-769, 1925.

1302 Lodge, Oliver J., Talks About Radio. With Some Pioneering History and Some Hints and Calculations for Radio Amateurs. London: Cassell; New York: George H. Doran, 1925. 267 pp. Pt. 1, pp. 17-119, on radio in general, is largely historical. A few refs. in the text.

1303 Marconi, Guglielmo, "Radio Dream Come True." Radio News, 11:784, 785, 849, 850, March 1930. 6 photos. Reminiscences from 1901 on long-distance transmission.

1304 Hammond, John H., and Ellison S. Purington, "A History of Some Foundations of Modern Radio-Electronic Technology." Proc. IRE., 45:1191-1208, Sept. 1957. 18 illus., most of them circuits. This important paper reveals early work done at the Hammond laboratory, primarily before and during World War I, most of it related to circuit developments for remote control by radio. Much of the material, from hitherto unpublished sources, concerns torpedoes, triode circuits, intermediate-frequency circuits, and frequency modulation techniques. 92 footnote refs., mainly to U.S. patents. Also, Espenschied, Lloyd, "Discussion of 'A History of Some Foundations of Modern Radio-Electronic Technology.' " Proc. IRE., 47:1253-1268, July 1959. 34 footnote refs. Includes rebuttal, replication and comments, with 35 additional footnotes. These papers contain highly personalized accounts of the background of some key radio researches, up to about 1930. They expose the student to the ramifications of radio-electronic history and provide a variety of valuable data. The contentious climate of the early decades of radio is brought to life by the critic's discursive remarks, 50 of which are dealt with individually in the rebuttal. See also 1346.

1305 Millikan, Robert A., Radio's Past and Future. Chicago: University of Chicago Press, 1931. A brief (15 pp.) nontechnical discussion of the history of radio.

1306 Lodge, Oliver J., "A Retrospect of Wireless Communication." Sci. Mon., 33:512-521, Dec. 1931. Also, Nature, 128:591, 592, Oct. 3, 1931. A quick glance at some highlights.

1307 Quäck, Erich, "Ten Years of Transradio--A Retrospect." Proc. IRE., 20:40-61, Jan. 1932. 21 photos, maps and diags., 3 tables. Description of transoceanic radiotelegraph progress on the Berlin-New York circuit and links to other countries, and development of shortwave operation.

1308 Wenstrom, William H., "Historical Review of Ultra-Short-Wave Progress." Proc. IRE., 20:95-112, Jan. 1932. 8 diags., 1 table. Primarily on the evolution of oscillator circuits; spark, regenerative, Barkhausen-Kurz, Gill-Morrell, magnetrons. 54 refs. from 1916. Also discussion, 21:315, 316, Feb. 1933.

1309 Bragdon, E. L., "Radio's Evolution." Radio News, 14:82, 83, 122, Aug. 1932. 10 illus. A brief review of developments over 14 years with mention of some earlier events.

1310 Terman, Frederick E., Radio Engineering. New York: McGraw-Hill, 1932. 688 pp. 418 illus., mainly diags. and graphs, 16 tables. 302 footnotes, many multiple entries, including numerous mathematical derivations and supplements to the text. This comprehensive engineering text deals with circuit fundamentals, vacuum tubes, amplifiers, oscillators, detectors, modulation, power sources, transmitters, receivers, antennas, propagation, radio measurements. There are also chapters on special tubes, radio aids to navigation, and sound equipment. A key book of the period.

1311 Ladner, Alan W., and Charles R. Stoner, Short Wave Wireless Communication Including Ultra-Short Waves. London: Chapman and Hall; New York: John Wiley, 1932. 2nd ed., 1934, 3rd ed., 1936, 4th ed., 1942. 573 pp., 342 illus., 151 chap. refs. Chap. 2, pp. 8-22, brief history, with chronological table, 42 entries from 1887. A standard text.

1312 DeSoto, Clinton B., Two Hundred Meters and Down. The Story of Amateur Radio. West Hartford, Conn.: The American Radio Relay League, 1936. 184 pp. After an introduction of the radio amateur, the book is divided into three parts; pioneers, development and recognition, and international high-frequency communication. There are many names, dates, call numbers, personal references and extracts in the text. No index.

1313 Gernsback, Hugo, "The Old E.I. Co. Days." Radio-Craft, 9:572-575, 630-640, March 1938. 45 small border illus. On the Electro Importing Company, founded by Gernsback in 1904, and radio history. Picture captions, p. 640.

1314 Lessing, Lawrence P., "Revolution in Radio," Fortune, 20:86-88, 116B, 119, 121, Oct. 1939. 4 photos. An informative article on E.H. Armstrong, his inventions and development of the first frequency modulation broadcast system. See also 380.

1315 Rolo, Charles J., Radio Goes to War. The Fourth Front. New York: George P. Putnam's Sons, 1942. 293 pp. A survey of wartime radio: American, British, French, German, Italian and Russian.

1316 Armstrong, Edwin H., "The New Radio Freedom." J.F.I., 232:213-216, Sept. 1941. Historical and philosophical remarks pertaining to FM radio.

1317 Rettenmeyer, Francis X., "Radio-Electronic Bibliography." Radio, pp. 26-28, 30, 32, Sept. 1942. 5, Amplification, Detection, Oscillators, Coils, 267 items from 1924. See 93.

1318 Terman, Frederick E., Radio Engineers' Handbook. New York: McGraw-Hill, 1943. 1019 pp., 869 illus., 88 tables. A well-known standard textbook on circuit elements and theory, vacuum tubes, amplifiers, oscillators, modulation and demodulation, power supplies, propagation, antennas, transmitters, receivers, navigational aids and measurements. Over 1050 footnotes, mostly refs., many multiple entries.

1319 Armstrong, Edwin H., "Vagaries and Elusiveness of Invention." Elect. Eng., 62:149-151, April 1943. Some philosophical thoughts and memories related to radio history.

1320 [U.S.]. "Air Force Radio Communications." Radio News, 29:56-59, 216-220, June 1943. Historical survey of radio developments in U.S. aviation from 1909. 5 photos.

1321 Rettenmeyer, Francis X., "Radio-Electronic Bibliography." Radio, 28:37, 38, 40, 42; 38, 40, 42, 44; 44, 46; 44, 46, 48, 50, Feb. to May 1944. 15, Household Receivers, 909 items from 1913. See 93.

1322 Rettenmeyer, Francis X., "Radio-Electronic Bibliography." Radio, 28:42, 44, 46, 48, 50, 52, 54, 56, 58, June 1944. 16, Transmitters, 366 items from 1913. See 93.

1323 de Forest, Lee, "Milestones in the Radio Industry." Radio News, 32:36, 37, July 1944. 7 illus. Also, Harbord, James G., "Radio Between Two Wars," pp. 39-42. 8 photos.

1324 Read, Oliver, "25 Years of Amateur Radio Progress." Radio News, 32:43-46, July 1944. 7 illus. Highlights of amateur radio from 1919.

1325 Armstrong, Edwin H., "Frequency Modulation--1922 and 1948." Radio-Craft, 19:20, 55, June 1948. 2 photos. Comments on the technical outlook of the 1920s and the prospects for FM radio broadcasting and receiver manufacturers.

1326 Bedford, Leslie H., "The Rise of Radio." J. Brit. IRE., 9:2-8, 1949.

1327 Armstrong, Edwin H., "Wrong Roads and Missed Chances--Some Ancient Radio History." Midwest Engineer, 3:3-5, 21, 25, March 1951. A study of the background of radio with reference to Marconi's early work and some problems of long-distance transmission.

1328 Tucker, D. G., "The History of the Homodyne and the Synchrodyne." J. Brit. IRE., 14:143-154, April 1954. Historical review, 62 refs. from 1922.

1329 Swinyard, William O., "The Development of the Art of Radio Receiving from the Early 1920's to the Present." Proc. IRE., 50:793-798, May 1962. 3 diags., 21 footnote refs. Although brief, this is a detailed and informative survey of receivers and allied developments.

1330 Noble, Daniel E., "The History of Land-Mobile Radio Communications." Proc. IRE., 50:1405-1414, May 1962. 5 illus. An informal story of mobile radio for police use in the United States from the early 1920s.

1331 "Fifty Years for the Advancement of Amateur Radio." RSGB Bull., 39:17-28, July 1963. 10 photos, list of the Society's presidents. A brief survey of amateur radio and the role of the Radio Society of Great Britain, founded July 1913 as the London Wireless Club.

1332 Schroeder, Peter B., Contact at Sea. Ridgewood, N.J.: Gregg, 1967. 149 pp., illus. On international maritime radio and its development, with a bibliography.

1333 Clarricoats, John, World at Their Fingertips. The Story of Amateur Radio in the United Kingdom and a History of the Radio Society of Great Britain. London: Radio Society of Great Britain, 1967. 307 pp., 32 plates and other illus. 2 tables and other lists. List of past presidents, honorary members, and vice-presidents of the Society, pp. 291-293. App., list of affiliated societies and clubs, 1921. A detailed record of people and events from the early years of radio. 169 footnotes, mainly supplementary material.

E. Circuits and Patents to 1920

Some of the more important technical papers and survey articles on original inventions are listed in this and the following section. Prominent patent landmarks include the development of the three-electrode tube and its circuits as an amplifier and oscillator, the superheterodyne, neutrodyne, and frequency modulation. For patents on electron tubes, see 949, 977, 979. Collected material on radio patents will be found in 1192, 1193, 1211, 1212, 1253, 1263, 1272, 1276, 1283, 1304, some of which are extracted in these two sections. References to U.K. and U.S. government publications and other patent literature are contained in some of the general works listed in Chap. 1, especially 1, 3, 12, 23.

1334 Marconi, Guglielmo, "Improvements in Transmitting Electrical Impulses and Signals, and in Apparatus Therefor." A History of Wireless Telegraphy, pp. 296-320. 14 diags. Reprint of the Provisional Specification, June 2, 1896. London: No. 12,039. See 1234.

1335 Fleming, John A., The Principles of Electric Wave

Telegraphy. App. III, pp. 638-641, list of 273 British patents grouped by year from 1896. See 1253.

1336 Llewellyn, Frederick B., "Birth of the Electron Tube Amplifier." Radio and Television News, 57:43-45, March 1957. 4 illus. Survey of events of 1912, primarily on the work of H.D. Arnold and Bell Telephone engineers.

1337 [Armstrong, E.H.]. "The Armstrong Patent." Radio Broadcast, 1:71, 72, May 1922. On the invention of the feedback circuit, with a photo of Armstrong and a reproduction of the signed and notarized diagram, Jan. 31, 1913.

1338 Williams, Henry S., "A Hook-Up for $500,000." Popular Radio, 3:218-222, March 1923. Two photos (Armstrong, Pupin), and original sketch of Armstrong's feedback circuit.

1339 Reisz, Eugen, "A New Method of Magnifying Electric Currents." Electrician, 72:721, 726-729, Feb. 6, 1914. 6 diags., 2 graphs. Also, 73:538, 539, July 3, 1914. Particulars of the Lieben-Reisz-Strauss gaseous triode tube and amplifier circuit.

1340 de Forest, Lee, "The Audion--Detector and Amplifier." Proc. IRE., 2:24-30, March 1914; reprinted IEEE Spectrum, 1:22-29, March 1964. Also, New York Herald, Nov. 7, 1913, p. 12; Wireless Age, 1(NS):273, Jan. 1914. On the first public demonstration of the Audion as a telephone amplifier. Includes a diagram of a two-stage cascade circuit and a photo of a three-stage amplifier.

1341 Meissner, Alexander, "The Audion as a Generator of High-Frequency Currents." Electrician, 73:702, July 31, 1914. Circuit diag., photo. Description of apparatus and experiment of March 1913.

1342 Armstrong, Edwin H., "Operating Features of the Audion." Includes 6 circuit diags. See 970.

1343 Hogan, John V. L., "Developments of the Heterodyne Receiver." Proc. IRE., 3:249-259, Sept. 1915. 8 diags. See also, 1:75-102, July 1913. 23 photos and diags. With notes and discussion.

1344 Armstrong, Edwin H., "Some Recent Developments

in the Audion Receiver." Proc. IRE., 3:215-247, Sept. 1915; reprinted Proc. IEEE., 51:1083-1097, Aug. 1963. 26 circuits, graphs and photos. Based on the Electrical World article (970), this extended paper was presented in New York on March 3, 1915. Includes discussion by de Forest and Armstrong that reveals the contentious atmosphere that perpetually surrounded these men. See: "Scanning the Issue," Proc. IEEE., 51:1082, Aug. 1963, for historical comments; also a letter by Benjamin F. Miessner, Proc. IEEE., 51:1798, Dec. 1963, for further comments on the question of prior invention.

1345 Miessner, Benjamin F., Radiodynamics. New York: D. Van Nostrand, 1916. 200 pp., 112 illus. An early book on remote control with details of experimental radio circuits. See next item.

1346 Miessner, Benjamin F., On the Early History of Radio Guidance. San Francisco: San Francisco Press, 1964. 86 pp., 22 illus., mostly of notebook entries and letters. App. 1, pp. 63-80, two excerpts from the appeal brief in the Splitdorf case. App. 2, pp. 81-84, typescript on "selective transmitter-receiver unit." A personal story of the inventor's work on remote control and early electron-tube circuits, particularly from 1912 to 1916, with much information concerning radio litigation, patents, priorities, and personalities. Numerous extracts from correspondence in the text. See 1304.

1347 Armstrong, Edwin H., "A Study of Heterodyne Amplification by the Electron Relay." Proc. IRE., 5:145-163, April 1917.

1348 Hull, Albert W., "The Dynatron--A Vacuum Tube Possessing Negative Electrical Resistance." Proc. IRE., 6:5-35, Feb. 1918; Wireless World, 6:148, 1918.

1349 Armstrong, Edwin H., "A New Method for the Reception of Weak Signals at Short Wave Lengths." Proc. Radio Club of America, pp. 1-6, Dec. 1919. 6 diags., including two with a list of components and values for practical construction. This pioneer paper on the superheterodyne was presented in New York on Dec. 3, 1919. Also, "A New System of Short Wave Amplification," Proc. IRE., 9:3-27, Feb. 1921. 6 diags.

1350 Fleming, John A., The Thermionic Valve and Its Developments in Radiotelegraphy and Telephony. There are many references to patents in the text and footnotes. Controversial issues between the author and Lee de Forest are discussed in Chap. 3, pp. 102-147, with references to patents as well as articles. See 977.

1351 Coursey, Philip R., Telephony Without Wires, pp. 367-396. The list of references has 197 entries containing one or more patents. See 1283.

1352 Houck, Harry W., "The Armstrong Super-Autodyne Amplifier." Radio Amateur News, 1:403-405, 439; 469-471, 508-510, Feb., March, 1920. 16 illus. First publication giving contructional details of the new superheterodyne receiver.

F. Circuits and Patents from 1921

1353 Armstrong, Edwin H., "The Regenerative Circuit." Electric Journal, 18:153, 154, April 1921. Reminiscences on the invention of the feedback circuit.

1354 Scott-Taggart, John, Thermionic Tubes in Radio Telegraphy and Telephony. There are numerous patent references in the text and footnotes. See 979.

1355 Armstrong, Edwin H., "Some Recent Developments of Regenerative Circuits." Proc. IRE., 10:244, Aug. 1922; "The Super-Regenerative Circuit," Proc. Radio Club of America, June 1922; "The Armstrong Super-Regenerative Circuit," Radio News, 4:239, 395, Aug. 1922. Photo of apparatus and 3 diags. Also Wireless World, 11:234, 1922.

1356 Brady, John B., "The Radio Patent Situation." Radio News, 4:850, 882, 884, 886, 890, 892, 894, 896, 898, 900, 902, Nov. 1922. 9 illus. Survey of manufacturers, inventions, devices, circuits, patents and litigation from 1884.

1357 Kesler, Charles H., "Famous Radio Patents." Radio Broadcast, 2:407-413, March 1923. 4 illus. On the de Forest and Armstrong regenerative circuits.

1358 Hazeltine, Louis A., "Tuned Radio-Frequency Amplification with Neutralization of Capacity Coupling." Proc. Radio Club of America, March 1923. Paper presented in New York, on March 2, on the Neutrodyne circuit (patented Aug. 1919), first used in the Navy SE-1420 radio receiver.

1359 Hazeltine, Louis A., "The Neutrodyne Receiver." Radio News, 4:1949, 2052, May 1923. 3 illus.

1360 Stark, Kimball H., "The 'Neutrodyne' Receiving System." Radio Broadcast, 3:38-41, May 1923. 6 illus.

1361 Ringel, Abraham, "The Super-Heterodyne." Wireless Age, 11:30-33, April 1924. 2 photos, 3 diags. Full report of E. H. Armstrong's paper presented on March 5 in New York. History, the heterodyne principle, results and practical design features. See 1362, 1366.

1362 [Armstrong, Edwin H.]. "Improvements on the Super-Heterodyne Receiver." Radio News, 5:1576, May 1924. 2 diags., 2 photos. Brief account of a talk given by Armstrong. The full paper was published in October, see 1366.

1363 Brady, John B., "The Vacuum Tube Patent Situation." Radio News, 5:1573, 1678-1685, May 1924. 3 diags.

1364 Arvin, W. B., "De Forest Now Controls Regeneration Patents." Radio News, 6:41, 129, 131, 132, July 1924. 2 illus. of original feedback circuits.

1365 Armstrong, Edwin H., "The Story of the Super-Heterodyne." Radio Broadcast, 5:198-207, July 1924. Its origin, developments and some recent improvements. 14 illus. of early receivers and circuits. Also, Proc. Radio Club of America, Feb. 1924.

1366 Armstrong, Edwin H., "The Super-Heterodyne--Its Origin, Development, and Some Recent Improvements." Proc. IRE., 12:539-552, Oct. 1924. Presented in New York, March 5, 1924. 13 illus. of early receivers and circuits.

1367 Edwards, Paul S., "The New Type of Superheterodyne." Popular Radio, 6:477-483, Nov. 1924. 5 photos of equipment and 3 circuit diags. An improved receiver developed by the U. S. Signal Corps.

1368 Brady, John B., "Regeneration and the Patent Situation." Radio News, 6:1424, 1545-1550, Feb. 1925. 15 illus. On the regenerative circuits of von Lieben, Armstrong, de Forest and others, and companies and litigation.

1369 Morse, A.H., "History of Radio Inventions." Radio News, 6:2048, 2049, 2188; 2236, 2237, 2278, 2280, 2282, 2284; 7:52, 53, 100; 184, 185, 210, 212; 296, 297, 358, 360, 362, 364, May to Sept. 1925. 29 illus. See 1192.

1370 May, Myra, "From Figures to Fame." Radio Broadcast, 7:451-457, Aug. 1925. 5 illus. On Louis A. Hazeltine and his Neutrodyne receiver.

1371 Morse, A.H., Radio: Beam and Broadcast. Its Story and Patents. The text, or abstracts, and numerous diagrams from 21 basic patents are included in the App., pp. 101-186. Other patent numbers and references are given in the main text or in footnotes. See 1192.

1372 Tyne, Gerald F.J., "The Saga of the Vacuum Tube." Many refs. relate to electron tube and circuit patents, mostly up to the mid-1920s. English translations not given. Many patent diagrams are also shown. See 949.

1373 Schottky, Walter, "On the Origin of the Super-Heterodyne Method." Proc. IRE., 14:695-698, Oct. 1926. 11 refs. to patents and papers from 1913. Historical note on technical matters related to the writer's early work and the Armstrong superheterodyne receiver.

1374 Flam, John, "The Maze of Radio Patents." Radio, 8:41, 42, 70, 72, Dec. 1926. Discussion of tuned radio-frequency, grid leak, regeneration, superheterodyne, vacuum tube and other patents.

1375 Information on U.S. Radio Patents and Suits. New York: Radio Manufacturers Assoc., 1926. 175 pp.

1376 Wright, Milton, "Successful Inventors--X." Sci. Am., 137:328, 329, Oct. 1927. On Louis A. Hazeltine and his Neutrodyne radio circuit. 4 illus.

1377 Blake, George G., History of Radio Telegraphy and Telephony. Reference list, pp. 353-403, contains numerous American, British and German patents among the 1125 entries. See 1193.

1378 Cocking, W.T., "The Evolution of the Superhet." Wireless World, 32:182, 183, March 1933. A brief review of progress over the last ten years.

1379 Guy, R. F., "FM and UHF. A Revealing Study of Its Early History, Present Uses, and Future Applications." Communications, 23:30, 32, 34-36, 79, 84, 85, Aug. 1943.

1380 Howe, G. W. O., "Early Radio Inventions; The Work of de Forest and Armstrong." Wireless Engineer, 20:521-523, Nov. 1943.

1381 McNicol, Donald, Radio's Conquest of Space. Patents are referred to in footnotes and in the text. See 1211.

1382 Armstrong, Edwin H., "A Study of the Operating Characteristics of the Ratio Detector and Its Place in Radio History." Proc. Radio Club of America, 25:3-20, Nov. 1948. 13 circuit diags., 2 graphs, 5 waveform pictures, 3 tables. 4 Apps., on limiter circuits, disclosure of the regenerative circuit early in 1914, demonstration ratio detector circuit, and relevant RCA documents. This important paper by the inventor of wide-band FM critically appraises limiting circuits and also covers much history from the early years. 22 footnotes.

1383 Maclaurin, William R., Invention and Innovation in the Radio Industry. App. 2, pp. 269-291, contains 16 separate tables that summarize patent data, particularly litigation, with names of inventors and companies, patent numbers, dates, disposition of suits and related matter. The extensive footnotes include many items on patents. See 1212.

1384 Armstrong, Edwin H., "Some Recent Developments in the Multiplexed Transmission of Frequency Modulated Broadcast Signals." Proc. Radio Club of America, 30:3-13, Oct. 1953. 5 circuit diags., 4 photos of equipment and a view of the antenna tower and station KE2XCC (formerly W2XMN) at Alpine, N. J. A historical survey from 1906 is followed by detailed discussion of the system problem, the receiver problem and field tests. A one-page appendix gives particulars of the demonstrations at Pupin Hall, Columbia University, October 13, 1953. 11 footnotes.

1385 Jewkes, John, David Sawers, and Richard Stillerman, The Sources of Invention. Case histories related to radio and electronics. See 327.

1386 Sturmey, S. G., The Economic Development of Radio. Patents, licences and related material are included in this survey of the British radio industry. See 1420.

1387 Howeth, Linwood S., History of Communications-Electronics in the United States Navy. App. A, pp. 513-546, Chronology from 64 B. C. to 1945, contains about 40 patent references from 1860. App. J, pp. 577-580, Radio Patents Owned by or Licensed to the United States Government, 1923, is a list of 146 patents (number, date of issue, name and title) from 1903. See 305.

G. Aircraft Radio and Navigational Aids

Entries in this section include original papers and bibliographies. See 1253, 1272 for the early history of directive antennas, also various titles in Sect. 9B. For related items pertaining to World War II, see 798, especially Vols. 2, 4.

1388 Braun, Karl F., "On Directed Wireless Telegraphy." Electrician, 57:222-224; 244-248, May 25, June 1, 1906. 19 diags., 5 tables. Description of antenna experiments. Also p. 220.

1389 Bellini, Ettore, and Alessandro Tosi, "A Directive system of Wireless Telegraphy." Elect. Eng. (London), 2:771-775, Nov. 1, 1907. Also editorial, p. 748.

1390 Round, Henry J., "Direction and Position Finding." J. IEE., 58:224, 1920. On work during World War I.

1391 Franklin, Charles S., "Short-Wave Directional Wireless Telegraphy." J. IEE., 60:930-938, 1922; Electrician, 88:593, 594, May 19, 1922; Wireless World, 10: 219-225, 1922. Review of experiments from 1916.

1392 Kolster, Frederick A., and Francis W. Dunmore, "The Radio Direction Finder and Its Application to Navigation." Washington: National Bureau of Standards, 1922. Scientific Paper No. 428.

1393 Keen, Ronald, Wireless Direction Finding. London:

Iliffe, 1938. 3rd ed., 803 pp., 549 photos and diags. Bibliography, pp. 759-787, 570 entries grouped by year from 1893. 1st ed., Direction and Position Finding by Wireless, 1922. 2nd ed., Wireless Direction Finding and Directional Reception, 1927, 490 pp. 4th ed., 1947, 1059 pp., over 700 refs. A standard work.

1394 Smith-Rose, Reginald L., and R.H. Barfield, A Discussion of the Practical Systems of Direction Finding by Reception. London: H.M.S.O., 1923. Radio Research Board, Special Report No. 1.

1395 Engel, F.H., and Francis W. Dunmore, "A Directive Type of Radio Beacon and Its Application to Navigation." Washington: National Bureau of Standards, 1923. Scientific Paper No. 480.

1396 Dellinger, John H., and Haradan Pratt, "Development of Radio Aids to Air Navigation." Proc. IRE., 16: 890-920, July 1928. 19 photos and diags. Development by Bureau of Standards of a radio beacon system (A-N radio range). 12 refs. from 1919.

1397 Joliffe, Charles B., and Elizabeth M. Zandonini, "Bibliography on Aircraft Radio." Proc. IRE., 16: 985-999, July 1928. Over 250 refs. from 1910 in 8 groups; including history, navigational aids, beacon systems and direction finders.

1398 Smith-Rose, Reginald L., "Radio Direction-Finding by Transmission and Reception: with Particular Reference to Its Application to Marine Navigation." Proc. IRE., 17:425-478, March 1929. A report on contemporary work with historical background. Over 100 refs. from 1907.

1399 Dellinger, John H., Harry Diamond, and Francis W. Dunmore, "Development of the Visual-Type Airway Radio Beacon." Proc. IRE., 18:796-839, May 1930. Summary of radio range development. 23 photos and diags. Bibliography, 33 items from 1907.

1400 Diamond, Harry, and Francis W. Dunmore, "A Radio Beacon and Receiving System for Blind Landing of Aircraft." Proc. IRE., 19:585-626, April 1931. 34 illus., 11 footnote refs.

1401 Tuska, Clarence D., "Radio in Navigation." J.F.I.,

228:433-443, 581-603, Oct., Nov. 1939. 8 diags., 2 maps. Early history, technical characteristics of directional radio, direction finders, automatic radio bearing indicators, radio beacons. 94 refs. from 1888, including 45 patents.

1402 Rettenmeyer, Francis X., "Radio-Electronic Bibliography." _Radio_, pp. 37-40, 44-46, May 1942. 1, Aviation Radio, 340 items in 8 groups, mainly from 1938. _See_ 93.

1403 Rettenmeyer, Francis X., "Radio-Electronic Bibliography." _Radio_, pp. 37-40, 42; 38, 40, 42, June, July 1943. 11, Direction Finding, 533 items from 1920. _See_ 93.

1404 Rettenmeyer, Francis X., "Radio-Electronic Bibliography." _Radio_, 28:41, 64, 66, 70, 72, 74; 58, 60, 62; 48, 70, 72; 42, 64, 66; 34, 35, 66; 50, 52, 54, 56; 63, 66, 68, 70, 72; 46, 48, 50; 49, 50, 60-63; 50, 51, Sept. to Dec. 1944, Jan. to June 1945. 17, Aircraft Radio, 1265 items from 1900. _See_ 93.

1405 Colin, Robert I., "Survey of Radio Navigational Aids." _Elect. Comm._, 24:219-261, June 1947.

1406 Smith, R. A., _Radio Aids to Navigation_. Cambridge: University Press, 1947. 114 pp. Description of radio altimeters, Loran, Shoran, Gee, Oboe, Consol and other systems.

1407 Pierce, J. A., "Electronic Aids to Navigation." _Advances in Electronics_, 1:425-451, 1948. 13 illus. Introduction, prewar methods, wartime developments, postwar proposals, range, accuracy. No refs.

1408 Diamond, Harry, "Radio Aids to Aviation." _Radio Engineering Handbook_ (1950), pp. 1068-1158. 40 diags., 3 tables. Description of U.S. government and commercial facilities and various navigational systems under general headings; radial, hyperbolic, circular, instrument landing, and integrated systems for traffic control, also radio altimeters and airborne installations. 81 refs. to books, papers and technical publications. Also "Aircraft Radio" in previous eds. _See_ 58.

1409 Briggs, Lyman J., "Early Work of the National Bureau

of Standards." Sci. Mon., 73:166-173, Sept. 1951. 7 illus. Includes aircraft radio range, blind landing guidance, the proximity fuse.

1410 Weike, Vernon I., "Fifty Years in Aeronautical Navigational Electronics." Proc. IRE., 50:658-663, May 1962. 3 diags., 2 footnotes.

1411 Howeth, Linwood S., History of Communications-Electronics in the United States Navy. Several chapters on aircraft radio and the radio direction finder. See 305.

1412 Colin, Robert I., "Otto Scheller and the Invention and Applications of the Radio-Range Principle." Elect. Comm., 40:359-368, 1965. 2 diags. from German patents of 1907 and 1916. 21 refs.

H. The Radio Industry

See Sect. 3G for industrial histories, especially 336-343; Sects. 5C, 5D, for company histories; Chap. 6 for broadcasting companies, especially 573, 576, 582, 593, 630, 642. Particulars of early commercial wireless telegraph companies are contained in 1263, various years. See also 1194, 1200, 1201, 1211, 1212.

1413 [U. S.]. Federal Trade Commission, Report on the Radio Industry. Washington: G. P. O., 1924. 347 pp. Development of the industry, patent agreements, practices concerning manufacture, sale and use of products. Much on RCA.

1414 Jome, Hiram L., Economics of the Radio Industry. Chicago: A. W. Shaw, 1925. 332 pp. Diags., tables, charts, map. Inventions, legal problems, growth, production and marketing.

1415 The Radio Industry. The Story of Its Development. As told by leaders of the industry to the students of the graduate school of business administration, George F. Baker Foundation, Harvard University. Chicago and New York: A. W. Shaw, 1928. 330 pp. Text of 11 lectures covering various aspects of radio; historical, technical, legal, merchandising, advertising. A broad record of contemporary status to 1927, with light technical treatment and great emphasis on American radio. 29 illus. showing sales literature of the early 1920s. No refs.

1416 [RCA]. "The Radio Industry." The Development of American Industries. 1st ed., pp. 763-800. Early history, status in 1912, high-power stations, rise of radio telephony and broadcasting, status in 1919, radio patent conflicts, development of broadcasting after 1921, American broadcasting companies, international radio communications, short-wave experiments, facsimile, RCA licenses, broadcast transmitter infringements. 4th ed., pp. 768-801, "The Radio and Television Industry," has additional material on research and development and the growth of television. See 337.

1417 Archer, Gleason L., Big Business and Radio. New York: The American Historical Company, 1939. 503 pp., 21 plates. A business history of the radio industry in the United States. There are frequent and extensive quotes from company records, agreements, contracts, licenses, court records, FCC testimony and correspondence of prominent people. The first 250 pages portray events from 1922 to 1925; the next 100 pages bring the story up to 1930. There are chapters on radio and talking pictures, radio and the industrial depression, radio broadcasting of today, television and facsimile with some history. Nontechnical. 80 footnote refs. A companion book to the author's History of Radio to 1926. See 1194.

1418 Weinberger, Julius, "Basic Economic Trends in the Radio Industry." Proc. IRE., 27:704-715, Nov. 1939. 7 graphs, 5 tables. Survey of American business; sales, tubes, broadcast receivers. 10 refs.

1419 Batcher, Ralph R., "Radio Industry Marks Twenty-Fifth Anniversary." Tele-Tech, 8:34-37, May 1949.

1420 Sturmey, S. G., The Economic Development of Radio. London: Gerald Duckworth, 1958. 284 pp., 37 footnotes, 7 tables. List of company titles (50), pp. 12, 13. A thorough study of inventions, innovations, companies and group activities in the British radio industry; marine and long-distance communications, broadcasting, development of radio valves, patents, wire relays. Much of historical interest, with many useful references to items in trade and technical magazines, government reports and newspapers. 346 chap. refs.

14. SOLID-STATE ELECTRONICS

Entries in this chapter concern crystalline solids, semiconductor theory, diodes, transistors, plate rectifiers, magnetic amplifiers, piezoelectricity, and other solid-state topics. They include original research papers, surveys, general articles, collections, bibliographies, and a few selected texts on semiconductor technology. Related material is contained in 661, 663 (magnetism); 742, 752, 755, 757, 759, 760, 767, 768, 799 (photoelectricity); 818, 822, 823, 826 (piezoelectricity); 1069 (components); 1086 (microelectronics); 1112 (computers).

A. Semiconductors to 1948

1421 Pierce, George W., "Crystal Rectifiers for Electric Currents and Electric Oscillations." Phys. Rev., 25: 31-60, July 1907; 28:153-187, March 1909; 29:478-484, Nov. 1909. Pt. 1, Carborundum; Pt. 2, Carborundum, Molybdenite, Anatase, Brookite; Pt. 3, Iron Pyrites. Diags., graphs, tables and waveform pictures.

1422 Pierce, George W., "Crystal Rectifiers." Principles of Wireless Telegraphy, pp. 157-200. 26 diags. and graphs, plate with 5 waveform pictures, 8 tables. Detectors by H. H. C. Dunwoody, L. W. Austin, and G. W. Pickard are described, with a full discussion of the author's experiments. See 1260.

1423 Eccles, William H., "The Energy Relations of Certain Detectors used in Wireless Telegraphy." Electrician, 66:166-168, Nov. 11, 1910. 12 graphs. Experiments and results with electrolytic, carborundum, zincite-chalcopyrite, and graphite-galena detectors. Also, "Electro-thermal Phenomena at the Contact of Two Conductors, with a Theory of a Class of Radiotelegraph Detectors." 71:900-903, Sept. 5, 1913. 6 diags.

1424 Eccles, William H., "Contact Detectors." Wireless

Telegraphy and Telephony, pp. 270-281. 14 graphs, 1 diag. Theory, characteristics, types of minerals used. See 1276.

1425 Pickard, Greenleaf W., "Oscillating Detector." QST, 4:44, March 1920.

1426 "The Crystodyne Principle." Radio News, 6:294, 259, 431, Sept. 1924. 3 diags., 2 photos. Experimental and receiving circuits employing an oscillating crystal.

1427 Podliasky, I., "Crystodyne Receivers and Amplifiers." Radio News, 6:470, 540, 542, Oct. 1924. 7 diags. Practical circuits employing a zincite oscillator.

1428 Pickard, Greenleaf W., "The Discovery of the Oscillating Crystal." Radio News, 6:1166, 1270, Jan. 1925. 5 illus. Historical account of the early work by Eccles and by the author in 1910.

1429 Lossev, Oles V., "Oscillating Crystals." Radio News, 6:1167, 1287-1291, Jan. 1925. 9 illus., 2 tables. Characteristics of oscillating crystals and circuits.

1430 Blake, George G., "Crystal Detectors." History of Radio Telegraphy and Telephony, pp. 87-91. See 1193.

1431 Bottom, Virgil E., "Invention of the Solid-State Amplifier." Physics Today, 17:24-26, Feb. 1964. On Julius E. Lilienfeld's solid-state patent of 1926. 3 diags., 4 refs. See also, J. B. Johnson, "More on the Solid-State Amplifier and Dr. Lilienfeld." Physics Today, 17: 60-62, May 1964, 15 refs. from 1897; R. W. Hall, "Invention of Solid-State Amplifier." IEEE Spectrum, 1:164, April 1964.

1432 Wilson, Alan H., "The Theory of Electronic Semiconductors." Proc. Roy. Soc., 133(A):458-491, Oct. 1, 1931; 134(A):277-287, Nov. 3, 1931.

1433 Wilson, Alan H., Semiconductors and Metals: an Introduction to the Electron Theory of Metals. Cambridge: University Press, 1939. 119 pp., diags. Chap. refs., also bibliographic note.

1434 Mott, N. F., "The Theory of Crystal Rectifiers."

Proc. Roy. Soc., 171(A):27-38, May 1, 1939.

1435 Scaff, J. H., and Russell S. Ohl, "Development of Silicon Crystal Rectifiers for Microwave Radar Receivers." B. S. T. J., 26:1-30, Jan. 1947. 18 diags., graphs, charts and photos, 2 tables. Point contact rectifiers, early microwave research, cartridge assemblies, shielded structures, types, applications, operating characteristics. 8 footnotes.

1436 Torrey, Henry C., and Charles A. Whitmer, Crystal Rectifiers. M. I. T. Rad. Lab. Series, Vol. 15, 1948. 443 pp., 215 photos, diags., graphs, 22 tables. Introduction, general properties, the crystal converter, special types, including fabrication, testing, applications. 290 footnotes and refs. See 798.

B. Transistors and Diodes from 1948

The references given in 1444, 1461, 1465, 1467, 1468, 1475, are particularly useful guides to specific topics.

1437 Bardeen, John, and Walter H. Brattain, "The Transistor, a Semi-conducting Triode." Phys. Rev., 74: 230, 231, July 15, 1948. 2 illus., 4 footnotes. Letter announcing the transistor. See also pp. 231-233.

1438 [Bell Laboratories]. "The Transistor." Bell Lab. Rec., 26:321-324, Aug. 1948.

1439 Fink, Donald G., and Frank H. Rockett, "The Transistor--A Crystal Triode." Electronics, 21:68-71, Sept. 1948. 4 illus.

1440 Rockett, Frank H., "The Transistor." Sci. Am., 179:52-55, Sept. 1948. 6 photos. Introduction to the germanium point contact transistor.

1441 Becker, Joseph A., and John N. Shive, "The Transistor--A New Semiconductor Amplifier." Elect. Eng., 68:215-221, March 1949. 12 diags. and graphs, 3 tables. Description of "the newly discovered device, the transistor." Construction, characteristics, performance, limitations, temperature effects. 9 refs. from 1947.

1442 Bardeen, John, and Walter H. Brattain, "Physical Principles Involved in Transistor Action." B. S. T. J., 28:239-277, April 1949. 15 illus. Two point contacts on germanium block. Theory, characteristics, and background of the development. Same, Phys. Rev., 75:1208-1225, April 1949. 48 refs. and notes from 1936.

1443 Ryder, Robert M., and R. J. Kircher, "Some Circuit Aspects of the Transistor." B. S. T. J., 28:367-400, July 1949. 30 illus. Amplifiers, equivalent circuit, Type A transistor, frequency response, power, distortion, noise. 7 refs. This issue (pp. 335-489) contains 4 other papers on transistors, and an editorial note.

1444 Shockley, William, Electrons and Holes in Semiconductors. With Applications to Transistor Electronics. New York: D. Van Nostrand, 1950. 558 pp., 1 plate, 142 illus., 10 tables. In 3 parts: 1, Introduction to transistor electronics; 2, Descriptive theory of semiconductors; 3, Quantum-mechanical foundations. Reference lists for each part, 72 entries, books and articles. Of some 320 footnotes, about half are references to published material. A pioneer text.

1445 Shive, John N., "The Phototransistor." Bell Lab. Rec., 28:337-342, Aug. 1950. See also, "A New Germanium Photoresistance Cell." Phys. Rev., 76:575, Aug. 15, 1949. 2 diags. 8 refs.

1446 [Bell Laboratories]. The Transistor; Selected Reference Material on Characteristics and Applications. New York: Bell Telephone Laboratories, 1951. 792 pp. Papers given at a symposium, Sept. 1951. Various topics; theory, properties, characteristics, circuit design, applications. Prepared for Western Electric Co. Includes some articles previously published in U. S. Illus. and refs.

1447 Scott, Thomas R., "Crystal Triodes." Proc. IEE., 98(Pt. 3):169-183, May 1951. Brief review of developments, with discussion. 30 refs. Also, Elect. Comm., 28:195-208, Sept. 1951.

1448 Ridenour, Louis N., "A Revolution in Electronics." Sci. Am., 185:13-17, Aug. 1951. 5 illus. On electron tubes, electronic equipment, and possibilities for transistors.

1449 "The Junction Transistor." Bell Lab. Rec., 29:379-381, Aug. 1951. 5 illus. Brief description of development and characteristics.

1450 Fink, Donald G., and R.K. Jurgen, "The Junction Transistor." Electronics, 24:82-85, Nov. 1951. 4 photos. Brief review of new design, improvements, advantages, characteristics. 5 refs.

1451 Armstrong, L.D., "Crystal Diodes in Modern Electronics." Radio Television News, 46:47-50, 162-166; 66-68; 60-62, 108, Oct.-Dec. 1951. 47:63-65, 127-133; 62-64, 120, 122, 123; 56, 57, 127, 128, Jan.-Mar. 1952. 59 illus., 5 tables. Basics of germanium diodes, theory, construction, manufacturing, application, circuits.

1452 Sparks, Morgan, "The Junction Transistor." Sci. Am., 187:28-32, July 1952. 10 illus.

1453 [Science Museum]. Semiconductors. London: Science Museum, Sept. 1952. 21 pp. Bibliography, 431 items from 1946.

1454 Slade, B.N., "Survey of Transistor Development." Radio Television News, 48:43-45, 170, 171; 64, 65, 112, 114-116; 68, 69, Sept.-Nov. 1952. 20 illus. Various footnote refs.

1455 [I.R.E.]. "Transistor Issue." Proc. IRE., 40:1283-1602, Nov. 1952. 51 papers plus introduction and editorial notices. A comprehensive collection on physics, theory, characteristics and applications. See next 2 entries.

1456 Morton, Jack A., "Present Status of Transistor Development." Proc. IRE., 40:1314-1326, Nov. 1952. 31 photos, graphs, diags. 14 footnote refs. Also, B.S.T.J., 31:411-442, May 1952. Reproducibility, reliability, miniaturization, performance, selected applications.

1457 Shockley, William, "A Unipolar 'Field Effect' Transistor." Proc. IRE., 40:1365-1376, Nov. 1952. 12 illus. Theoretical paper. 9 refs.

1458 Bello, Francis, "The Year of the Transistor." Fortune, 47:128-133, 162, 164, 166, 168, March 1953. 7 photos, 6 diags. Survey of electronic technology; computers, television, hearing aids, with a pictorial guide to transistors for the layman. Some historical background.

1459 Bradley, William E., and others, "The Surface Barrier Transistor." Proc. IRE., 41:1702-1720, Dec. 1953. 5 papers, illus. 1, Principles; 2, Fabrication; 3, Circuits; 4, High-frequency performance; 5, Properties of metal to semiconductor contacts. Prepared by Technical Staff, Philco Research Division. Various refs.

1460 Barton, Loy E., "An Experimental Transistor Personal Broadcast Receiver." Proc. IRE., 42:1062-1066, July 1954. 8 diags. and photos, table. Laboratory model AM receiver with 9 alloy-junction transistors and 2 diodes.

1461 Krull, Alan R. (comp.), Transistors and Their Applications; a Bibliography, 1948-1953. Evanston, Ill.: Northwestern University, Technological Institute Library, 1954. 77 pp. Over 900 refs. to world literature, with some titles from 1938. Also, Trans. IRE., ED-1:40-70, Aug. 1954.

1462 Bown, Ralph, "The Transistor as an Industrial Research Episode." Sci. Mon., 80:40-46, Jan. 1955. 8 diags. and photos. A personal story of transistor research and development within Bell Telephone Laboratories. 2 refs.

1463 Herold, Edward W., "Semiconductors and the Transistor." J.F.I., 259:87-106, Feb. 1955. 14 illus., 2 tables. A survey of history, semiconductors, modern techniques, present commercial devices and laboratory developments. 33 refs. from 1874.

1464 Dacey, George C., and Ian M. Ross, "Field Effect Transistor." B.S.T.J., 34:1149-1189, Nov. 1955. 22 illus., 3 tables. Theory and experimental results with a discussion of fabrication and performance. 4 refs.

1465 Pearson, Gerald L., and Walter H. Brattain, "History of Semiconductor Research." Proc. IRE., 43:1794-1806, Dec. 1955. 2 tables. A comprehensive story of semiconductor research over 120 years; point-contact, cuprous oxide, and selenium rectifiers, radar diodes, transistors--theories, experiments, materials. Bibliography, 118 articles, patents and books, from 1839.

1466 Scott, Thomas R., Transistors and Other Crystal Valves. London: Macdonald and Evans, 1955. 258 pp., 66 illus., including plates, 26 tables. Bibliography, pp. 237-253, 277 entries, some multiple, from 1835. Brief historical survey, pp. 1-7, with 26 refs.

1467 Hunter, Lloyd P. (ed.), Handbook of Semiconductor Electronics. New York: McGraw-Hill, 1956. Over 600 pp., 484 illus., 37 tables. "A practical manual covering the physics, technology, and circuit applications of transistors, diodes, and photocells." 20 sections by 14 authors. List of symbols, pp. xix-xxix. 313 chap. refs. Bibliography (comp. Wayne A. Kalenick), 68 pp., over 2000 entries grouped by year (from 1936) and alphabetically by author. A comprehensive handbook with historically useful refs. 2nd ed., 1962, 900 pp., 496 illus.

1468 Garner, Louis E., Transistor Circuit Handbook. Chicago: Coyne Electrical School, 1956. 420 pp., 192 illus. Basic circuits, practical applications and transistor data. Bibliography, pp. 397-410, 221 entries in 7 groups: Lists, books, trade publications, general periodicals, magazine articles, technical papers, patents.

1469 [RCA]. Transistors, 1. Princeton, N. J.: R. C. A. Laboratories, 1956. 676 pp. 41 papers by RCA authors, 46 abstracts. Over 200 refs.

1470 Esaki, Leo, "New Phenomenon in Narrow Germanium p-n Junctions." Phys. Rev., 109:603, 604, Jan. 15, 1958. 2 graphs, diag.

1471 Ryder, Robert M., "Ten Years of Transistors." Radio-Electronics, 29:34-37, May 1958. 4 diags., 4 photos. Condensed history with highlights of progress. 11 refs.

1472 [I. R. E.]. "Transistor Issue." Proc. IRE., 46:949-1360, June 1958. 43 papers covering a broad range of topics. Extensive refs. See next entry.

1473 Morton, Jack A., and William J. Pietenpol, "The Technological Impact of Transistors." Proc. IRE., 46:955-959, June 1958. 2 graphs.

1474 Baker, W. O., "The First Ten Years of the Transistor." B. S. T. J., 37:i-vi, Sept. 1958. A guest editorial on the broader view of semiconductor technology.

1475 Meyrick, N. L. (comp.), Fifteen Years of Semi-Conducting Materials and Transistors: a Bibliography. Newmarket, England: Newmarket Transistor Co., 1958. Over 2500 items from 1941.

1476 Early, James M., "Semiconductor Devices." Proc. IRE., 50:1006-1010, May 1962. Brief review of history, technical influence, status and possibilities. 10 refs.

1477 Petritz, Richard L., "Contributions of Materials Technology to Semiconductor Devices." Proc. IRE., 50: 1025-1038, May 1962. 19 illus., 2 tables. Research on germanium and silicon, and process technology. 76 refs. from 1917.

1478 Wallmark, J. Torkel, "The Field-Effect Transistor--A Review." RCA Rev., 29:641-660, Dec. 1963. 16 illus. Chronological bibliography, 54 items from 1939. Also, "The Field-Effect Transistor--An Old Device with New Promise." IEEE Spectrum, 1:182-191, March 1964. 15 illus., 2 tables.

1479 Kawakami, Masamitsu, and Kiyoshi Takahashi, "The Evolution of Semiconductor Electronics." Electronic Industries, 24:72, 73, Feb. 1965. Brief text and a "tree" chart showing genealogy.

1480 "The Transistor: Two Decades of Progress." Electronics, 41:77-130, Feb. 19, 1968. 35 illus. Staff report with other contributors. Trade review.

1481 Costigan, Daniel M., "The Quest for the Crystal that Amplifies." Popular Electronics, 28:40-44, June 1968. 3 photos. A brief history.

C. Other Solid-State Devices

This section contains miscellaneous solid-state topics; properties and applications of materials and devices, magnetic amplifiers, plate rectifiers, thermistors, varistors, and piezoelectricity. Relevant items will be found in other Sects., 3A, 7A, 7E, 7G, 8A, 9B, 11A, 11C, 11E. See 194, 662, 804, 817, 818, 822, 823, 827, 922, 1049, 1069, 1083. Several volumes in the Radiation Laboratory Series (798) also contain related material.

1482 Alexanderson, Ernst F.W., "A Magnetic Amplifier for Radio Telephony." Proc. IRE., 4:101-120, April 1916. First report of practical application.

1483 Grondahl, Lars O., and P.H. Geiger, "A New Electronic Rectifier." Trans. AIEE., 46:357-366, Feb. 1927. 18 photos, diags., graphs and waveforms. Description of the copper-oxide rectifier; characteristics, design, theory, applications. 5 footnotes. Also, Grondahl, L.O., "The Copper-Cuprous-Oxide Rectifier and Photoelectric Cell." Rev. Mod. Phys., 5:141-168, April 1933. 23 photos, graphs, diags., 3 tables. Bibliography, 141 items from 1874, grouped under rectifiers, photoelectric cells, properties of cuprous oxide.

1484 Becker, Joseph A., "Varistors--Their Characteristics and Uses." Bell Lab. Rec., 18:322-327, July 1940. Rectifiers, symmetrical varistors, thermistors, types, applications.

1485 Williams, A.L., and L.E. Thompson, "Metal Rectifier." J. IEE., 88(Pt. 1): 353-383, Oct. 1941. A survey of copper oxide and selenium rectifiers, with a history of developments. 130 refs. from 1874.

1486 Rettenmeyer, Francis X., "Radio-Electronic Bibliography." Radio, pp. 37-43, July 1942. 3, Crystallography, 284 items from 1811. See 93.

1487 Cady, Walter G., Piezoelectricity. New York: McGraw-Hill, 1946. 806 pp., 168 illus., 36 tables. Refs. with most chaps., numerous footnotes. General bibliography, pp. 755-787, lists 57 books and 602 articles from 1894, with periodical names and abbreviations. This comprehensive treatise is a standard work.

1488 Becker, Joseph A., C.B. Green, and G.L. Pearson, "Properties and Uses of Thermistors, Thermally Sensitive Resistors." B.S.T.J., 26:170-212, Jan. 1947. 25 diags., graphs, photos, 3 tables. Materials, structure, processes, properties of semiconductors, physical properties of thermistors, characteristics, applications. 15 refs. from 1928. Also, Trans. AIEE., 65:711-725, Nov. 1946.

1489 Grondahl, Lars O., "Twenty-five Years of Copper Oxide Rectifiers." Trans. AIEE., 67:403-410, 1948.

History from 1920, development, applications, manufacturing, testing.

1490 [Westinghouse Electric Corporation]. "Twenty-One Years of Metal Rectifiers." Engineer, 186:319, Sept. 24, 1948; Engineering, 166:331, 332, Oct. 1, 1948. A survey of the history of developments of copper oxide and selenium rectifiers, based on a Westinghouse exhibition.

1491 Henisch, Heinz K., Metal Rectifiers. Oxford: Clarendon Press, 1949. 155 pp. Contains a comprehensive bibliography of 556 items from 1835.

1492 Miles, J. C., "Bibliography of Magnetic Amplifier Devices and the Saturable Reactor Art." Trans. AIEE., 70(Pt. 2):2104-2123, 1951. 900 books, articles and patents from 1887.

1493 Hogan, C. Lester, "The Ferromagnetic Faraday Effect at Microwave Frequencies and Its Applications: The Microwave Gyrator." B. S. T. J., 31:1-31, Jan. 1952. 14 illus., table. Historical background, theory, equipment and measuring techniques, experimental results, materials, applications. 12 refs.

1494 Coales, J. F., "Magnetic Amplifiers." Proc. IEE., 101(Pt. 2):83-99, April 1953. Historical review is included. 20 refs. from 1908.

1495 Geyger, William A., Magnetic-Amplifier Circuits: Basic Principles, Characteristics and Applications. New York: McGraw-Hill, 1954. 2nd ed., 1957, 394 pp. Includes history. Extensive chap. refs., including patents, over 1200 items.

1496 Lufcy, Carroll W., "A Survey of Magnetic Amplifiers." Proc. IRE., 43:404-413, April 1955. 18 illus. Principles, circuits, theory, components, materials, applications. Includes a brief historical review. 26 footnote refs.

1497 [I. R. E.]. "Solid-State Electronics Issue." Proc. IRE., 43:1701-1973, Dec. 1955. 17 papers. Silicon and germanium, crystal phosphors, ferroelectricity, photoconductors, electroluminescence. One historical paper is listed separately (1465).

1498 [I. R. E.]. "Ferrites Issue." Proc. IRE., 44:1233-

1468, Oct. 1956. 27 papers. Theory, materials and properties, devices and applications. See next entry.

1499 Owens, C. Dale, "A Survey of the Properties and Applications of Ferrites Below Microwave Frequencies." Proc. IRE., 44:1234-1248, Oct. 1956. 19 illus., 2 tables. 70 refs. arranged by year from 1946.

1500 [A. I. E. E.]. "Magnetic Amplifier Bibliography, 1951-1956." Communications and Electronics, No. 39, pp. 613-627, Nov. 1958. Covers world literature and patents. Chronologically arranged in 4 sections, subdivided by topic. 849 items. Includes list of journals.

1501 Henisch, Heinz K., Rectifying Semi-Conducting Contacts. Oxford: Clarendon Press, 1957. 372 pp., 129 illus., 3 tables. A comprehensive study with historical notes, pp. 7-18. Refs., pp. 352-367, 642 entries grouped by year from 1874.

1502 Katz, Harold W. (ed.), Solid State Magnetic and Dielectric Devices. New York: John Wiley, 1959. 542 pp., illus. A thorough survey of materials, components, devices, applications. Chap. refs., 470 items from 1858.

1503 Button, Kenneth J., "Historical Sketch of Ferrites and Their Microwave Applications." Microwave J., 3:73-79, March 1960. Review of developments with 82 refs. from 1896.

1504 Mumford, William W., "Some Notes on the History of Parametric Transducers." Proc. IRE., 48:848-853, May 1960. A review of developments from 1831, with 200 refs. from that year in chronological order.

1505 Bottom, Virgil E., "Piezoelectric Effect and Applications in Electrical Communication." Proc. IRE., 50:929-931, May 1962. 2 illus. A short survey from 1880. 12 refs.

1506 Lax, Benjamin, and John G. Mavroides, "Solid-State Devices Other than Semiconductors." Proc. IRE., 50:1011-1024, May 1962. 15 illus. Historical survey including ferromagnetic, electromechanical and microwave devices, dielectric materials, computer elements, superconductors, masers. About 100 refs. from 1847, in 84 entries, plus 9 other refs.

15. TELEGRAPHY AND TELEPHONY

Electrical engineering began with the advent of the wire telegraph around 1840. By 1880, the art had expanded to include the telephone and power machinery. The laying of the Atlantic cables (1857-1866) was the high point of engineering achievement during these forty years. Stories of the telegraph and the telephone abound throughout technical and popular literature, therefore many titles listed elsewhere are pertinent, especially the following. Catalogs, 7, 12, 15. Electricity, 238, 239, 247, 250, 251. Electrical engineering, 259, 261, 273, 274. Telecommunications, 275, 276, 287, 308. Inventors and Inventions, 312, 314, 321. Industries, 332, 336, 337. Collected biographies, 359, 360, 368, 369. Individual biographies, 385-388, 398-400, 414-416, 443, 444. Commercial organizations, 522-528, 534-539, 543, 554-558. Radio, 1234, 1249. Extracts from some of these items are listed in this chapter.

A. The Telegraph

1507 Shaffner, Taliaferro P., The Telegraph Manual: A Complete History and Description of the Semaphoric, Electric and Magnetic Telegraphs of Europe, Asia, Africa, and America, Ancient and Modern. New York: Pudney and Russell, 1859. 850 pp., 10 plates (ports.), 625 illus. This comprehensive survey includes semaphore telegraphs, principles of electricity and magnetism, early electric telegraphs, details of the prominent systems in each major country, submarine telegraphs, telegraph construction and maintenance, organization and administration of national systems. Instruments, apparatus, poles, wires, insulators, cables, installation methods, and system operations are described in detail. App., pp. 803-844, contains biographical sketches of ten eminent telegraphers. The basic source for early telegraph history.

1508 Noad, Henry M., "The Electric Telegraph." A Manual of Electricity, pp. 747-803. 42 illus. Early history and detailed descriptions of instruments and systems, including electric clocks. See 644.

1509 Prescott, George B., History, Theory, and Practice of the Electric Telegraph. Boston: Ticknor and Fields, 1860. 3rd ed., 1866. 508 pp., 111 engrs. Electricity and magnetism, general principles of the electric telegraph, submarine cables and the Atlantic cable, telegraph instruments and apparatus, line construction, and practical working. Various systems are described in detail; needle, Morse, Bain, Hughes, dial, printing, and others. There is much historical background on the invention and progress of the electric telegraph, "the foremost invention of the age."

1510 Culley, R. S., A Handbook of Practical Telegraphy. London: Longmans, Green, 1863. 191 pp., text illus. 7th ed., 1878. 468 pp., 15 plates, some folding, 147 diags. A detailed treatment of electrical and magnetic theory, telegraph apparatus, line construction and submarine cables. Apparatus and methods for testing and fault-finding also treated. A notice of recent inventions (7th ed.) includes the telephone and the quadruplex system.

1511 Thompson, Robert L., Wiring a Continent. The History of the Telegraph Industry in the United States, 1832-1866. Princeton, N.J.: Princeton University Press, 1947. 544 pp., 24 plates, 11 maps, 7 text illus., folding chart. Extensively documented with 1239 footnotes, mainly refs. to original papers, correspondence, company records and court documents. 15 Apps., pp. 447-517, various agreements and contracts. Bibliography, pp. 518-526, manuscript sources, 16 entries; books and articles, 140 entries. Primarily a business history.

1512 Dodd, George, Railways, Steamers, and Telegraphs; a Glance at Their Recent Progress and Present State. London, Edinburgh: W. and R. Chambers, 1867. 326 pp., 10 plates. This is an early work on the unity between the new modes of land and sea transportation and the electric telegraph. Chap. 3, pp. 219-319, Telegraphs. Mainly on the Atlantic cables. Chronological list, pp. 320-322, from 1815 to 1866.

1513 Sabine, Robert, The History and Progress of the Electric Telegraph. With Descriptions of Some of the Apparatus. London: Virtue; New York: Virtue and Yorston, 1867. 2nd ed., D. Van Nostrand, 1869. 280 pp., 134 illus. 3rd ed., London: Lockwood, 1872. 280 pp., text figures and folding plate.

1514 Pope, Franklin L., Modern Practice of the Electric Telegraph. A Handbook for Electricians and Operators. New York: Russell Brothers, 1869, 128 pp. D. Van Nostrand, 1872. 6th ed., revised and enlarged. 160 pp., 65 engrs., 4 plates, showing instruments, apparatus and circuit diagrams. Batteries, circuits, the Morse system, insulators, testing, construction, hints to learners, recent improvements, technical notes. 10th ed., 1877, 160 pp. 12th ed., 1888; 14th ed., 1891, 234 pp., 185 illus.

1515 Sharlin, Harold I., "The Telegraph." The Making of the Electrical Age, pp. 7-35. Brief history to 1872. 54 refs. See 250.

1516 Preece, William H., and James Sivewright, Telegraphy. London: Longmans, Green, 1876. 300 pp., illus. A standard text. Many later eds.

1517 Prescott, George B., Electricity and the Electric Telegraph. New York: D. Appleton, 1877. 978 pp., illus., tables and folding frontispiece. 4th ed., 1881. 6th ed., revised and enlarged, 2 vols., 1885, 670 illus.

1518 Guillemin, Amédée, The Applications of Physical Forces, pp. 543-632. Electric telegraphy and telegraphic lines. History, general theory; needle, dial, alphabetic, printing, and autographic telegraphs. Air and subterranean lines, submarine cables, galvanometers, batteries, alarums, duplex telegraphy. There are 66 engrs. of instruments and cables, and one plate showing Hughes' printing telegraph. Also Chap. 6, pp. 633-650, on electric horology; clocks, regulators, time signals and chronoscopes (11 illus.). See 191.

1519 Reid, James D., The Telegraph in America: Its Founders, Promoters and Noted Men. New York: Derby Brothers, 1879. 846 pp., illus. History of notable inventors and telegraph companies.

1520 Johnston, William J., Telegraphic Tales and Telegraphic History. New York: The Author, 1880. 254 pp. 2nd ed., 1882, 286 pp. According to the subtitle, a popular account of the electric telegraph, its uses, extent and outgrowths.

1521 Benjamin, Park, "The Electric Telegraph." The Age of Electricity, pp. 208-271. History and survey of various systems. See 237.

1522 Preece, William H., "Fifty Years' Progress in Telegraphy. The Jubilee of the Electric Telegraph." Electrician, 19:242-247, 249-253, July 29, 1887. In celebration of the experiment between Euston and Camden Town, July 25, 1837. Also, pp. 248, 249, "The Telegraph Jubilee Banquet."

1523 Walmsley, Robert M., "The Electric Telegraph." Electricity in the Service of Man, pp. 786-876. 92 illus. History and modern systems. Full description of instruments, installations and circuits. See 238.

1524 Buckingham, Charles L., "The Telegraph of Today." Electricity in Daily Life, pp. 138-172. Beginnings, descriptions of various systems, including cables. See 239.

1525 Munro, John, Heroes of the Telegraph. London: Religious Tract Society, 1891. 288 pp. Popular story with biographies.

1526 Maver, William, American Telegraphy and Encyclopedia of the Telegraph. New York: Maver Publishing Company, 1892. 5th ed., 1909. 668 pp., 430 illus. A full treatment of all kinds of systems and apparatus, including writing telegraphs, time services, ticker systems, burglar and fire alarms, police and railway signaling. Descriptions of individual instruments are given in great detail and supported by illustrations of mechanical parts, construction, assembly, operation and circuits.

1527 Routledge, Robert, The Telegraph. London: George Routledge, 1895. 289 pp., 193 illus. Trans. from the French work of A.-L. Ternant. Pt. 1, pp. 4-77, optical, acoustic, pneumatic systems. The remainder treats the electric telegraph; history, land lines, submarine cables, apparatus, applications. No index.

1528 Fahie, John J., A History of Wireless Telegraphy. Most of the book is concerned with conduction and induction methods of signaling without intervening wires. First period--the possible, pp. 1-78, from 1838 to 1872. Second period--the practicable, pp. 79-135, from 1880 to 1894. Preece's and Smith's methods, pp. 136-176. See 1234.

1529 Story, Alfred T., The Story of Wireless Telegraphy. Telegraph history with emphasis on conduction and induction methods, pp. 9-100. See 1249.

1530 Herbert, T. E., Telegraphy. A Detailed Exposition of the Telegraph System of the British Post Office. London: Pitman, 1906. 5th ed., 1930. 1199 pp., 735 illus. References to original papers are given in numerous footnotes. A standard text.

1531 Maver, William, and Donald McNicol, "American Telegraph Engineering--Notes on History and Practice." Trans. AIEE., 29:1303-1339, discussion pp. 1339-1356, 1910. 12 illus. History from 1840, sources of emf, printers, repeaters, superimposed systems, inductive disturbances, relays, engineering details, the telephone in testing, line construction.

1532 Kempe, Harry R., and Emile Garcke, "Telegraph." Encyclopaedia Britannica, 1911. 11th ed., 26:510-529. 34 illus. Land and submarine telegraphy, instruments, apparatus, cables and lines. History and commercial aspects, table of statistics from 1870 to 1907. Bibliography.

1533 Harrison, H. H., "The Story of Land Telegraphy." Electrician, 77:798-800, Sept. 15, 1916.

1534 Harrison, H. H., "Sixty Years of Telegraph Progress." Electrician, 87:600, 601, Nov. 11, 1921. This brief survey mentions wireless telegraphy.

1535 Roberts, A. H., "The Romance and History of the Electric Telegraph." J. POEE., 16:121-130, 207-224, 301-321; 17:1-19, 91-96, 171-180, 1923-1925.

1536 [Western Union Telegraph Company]. "The Telegraph Industry." The Development of American Industries, pp. 687-719. Early history, Morse, the Atlantic cables, Western Union, Postal Telegraph and Cable Company, legislation, economic significance, commercial development, the telegraph system of today (1931). Compare other eds. See 337.

1537 Bibber, H. W., "The Centenary of the Morse Telegraph." Elect. Eng., 63:433-436, Dec. 1944. Photo. 5 refs.

1538 D'Humy, Ferdinand E., and P. J. Howe, "American Telegraphy After 100 Years." Trans. AIEE., 63: 1014-1032, Dec. 1944.

1539 Hillis, George C., "Telegraphy--Pony Express to Beam Radio." Smithsonian Institution, Annual Report for 1947, pp. 191-205 (Washington, 1948). Map of microwave relay system, 3 plates, 4 refs. Also, J. Western Soc. Engineers, Vol. 51, No. 3, Sept. 1946.

1540 Coggeshall, Ivan S., "Telegraphy's Next 25 Years." Elect. Eng., 78:493-499, May 1959. 5 photos. A two-way view of progress and future trends.

B. Submarine Telegraphs and Cables

1541 Briggs, Charles F., and Augustus Maverick, The Story of the Telegraph, and a History of the Great Atlantic Cable. New York: Rudd and Carleton, 1858. 255 pp., 14 illus., port., folding map. 10 Apps., various topics. Early telegraph history, origins of the Atlantic telegraph, cable construction and experiments, expeditions of 1857 and 1858. No index.

1542 Field, Henry M., History of the Atlantic Telegraph. New York: Charles Scribner, 1866. 364 pp., 3 illus. A detailed account from 1850 to the successful attempt of 1866. No index. 2nd ed., 1892, 415 pp., 15 plates, 5 illus.

1543 Russell, William H., The Atlantic Telegraph. London: Day and Son (1866). 117 pp., 28 tinted lithograph plates. Laying of the first cable from Valentia to Hearts Content in 1858, the disasters of this and following cables, the various attempts, and the role of the Great Eastern as a special cable-laying steamship. The plates illustrate ships and cable-laying scenes, ship machinery on deck and within the cable holds, and shore stations. An authentic account and a basic source.

1544 McDonald, Philip B., A Saga of the Seas. The Story of Cyrus W. Field and the Laying of the First Atlantic Cable. New York: Wilson-Erickson, 1937. 288 pp., 12 plates. Bibliography, pp. 281, 282, 21 entries.

1545 Dibner, Bern, The Atlantic Cable. Norwalk, Conn.:

Burndy Library, 1959. 96 pp., 52 illus. and folding map. Apps., pp. 89-93, letters by Morse and Whitehouse. Bibliography, pp. 94, 95, 61 books and pamphlets in the Burndy Library. 39 footnotes. Also, Elect. Eng., 78: 892-898, 1014-1019, 1106-1111, Sept.-Nov. 1959.

1546 Marland, Edward A., "Submarine Telegraphs." Early Electrical Communication, pp. 153-172. Also "Submarine Signalling," pp. 173-181. History to 1866. 52+ 18 footnotes. See 275.

1547 Smith, Willoughby, The Rise and Extension of Submarine Telegraphy. London: Virtue, 1891. 390 pp., 6 plates, illus.

1548 Bright, Charles, Submarine Telegraphs; their History, Construction and Working. London: Crosby, Lockwood, 1898. 744 pp., 38 plates, 2 folding maps, diags.

1549 Bright, Edward B., and Charles Bright, The Life Story of the Late Sir Charles Tilston Bright, Civil Engineer. With which is incorporated the story of the Atlantic cable, and the first telegraph to India and the Colonies. London: Constable, 1899, 2 vols. Vol. 1, 506 pp., 81 illus., including plates, 3 folded maps and charts. 12 Apps., principally on the Atlantic telegraph, including two official reports (1857, 1858). Vol. 2, 692 pp., 84 illus., including plates. 27 Apps. include papers, reports, addresses and newspaper items. A comprehensive record put together by his brother and son. Revised and abridged (Charles Bright, ed.), 1908, 478 pp.

1550 Bright, Charles, The Story of the Atlantic Cable. London: George Newnes; New York: D. Appleton, 1903. 220 pp., 46 illus. Evolution of Atlantic telegraphy, making and laying the lines, commercial operation, and cable systems of today. No index. See also, Electrician, 77:801-807, Sept. 15, 1916. 15 photos, ports., diags.

1551 Garnham, S. A., and Robert L. Hadfield, The Submarine Cable; the Story of the Submarine Telegraph Cable from Its Invention down to Modern Times. London: Low, Marston, 1934. 242 pp., plates and diags. Cable history, cables and systems, cable ships, operation and service.

1552 "Centenary of Submarine Telegraphy." Engineering, 170:201, 202, Sept. 1, 1950. Editorial remarks on history and the exhibit at the Science Museum, London.

1553 Garratt, Gerald R. M., One Hundred Years of Submarine Cables. London: H. M. S. O., 1950. 59 pp., 16 plates with 28 illus. and a cable map of the world. Almost wholly on British achievements. Early years of land telegraphy, early submarine cables, first Atlantic cables, gutta percha submarine cables, growth of the long-distance cable network from 1870 to 1920, British network from 1920 to 1950, development of cables for submarine telephony and high-speed telegraphy. Bibliography, p. 57, 24 items from 1836 to 1939. See also 222, Vol. 4, Chap. 22.

1554 Kelly, Mervin J., and others, "A Transatlantic Telephone Cable." Communication and Electronics, No. 17:124-139, March 1955. 16 photos, maps and diags. Early telegraph cables, radio circuits, early American work, British experience. The transatlantic telephone cable project, basic features, general description, details of the cables, repeaters, testing methods and cable-laying plans. Discussion, pp. 136-139. A general survey with technical details of a plan to link the United Kingdom, Canada and the United States with the first cable telephone system, completed in 1956. 10 refs. from 1947. Also, Elect. Eng., 74:192-197, March 1955. A brief version with 5 illus., no refs. See also, B. S. T. J., 36:1-326, Jan. 1957. 11 papers on the transatlantic telephone cables.

1555 Clarke, Arthur C., Voice Across the Sea. London: Frederick Muller; New York: Harper, 1958. 220 pp., 12 plates. General account of submarine cables and communications from early telegraph days.

1556 Haigh, K. R., Cableships and Submarine Cables. Washington: U. S. Undersea Cable Corp., 1968. 405 pp., illus. History of submarine cable laying from 1850 up to the present. Documented.

C. The Telephone

1557 Watson, Thomas A., The Birth and Babyhood of the Telephone. New York: AT&T, 1913, 1938. 47 pp., 30 illus., ports., apparatus, views and maps. Address before the Third Annual Convention of the Telephone Pioneers of America, Chicago, Oct. 17, 1913.

1558 Watson, Thomas A., "How Bell Invented the Telephone." Proc. AIEE., 34:1503-1513, Aug. 1915; Elect. Eng., 66:232-236, March 1947. 2 illus. A personal account of the events of 1875-1877.

1559 Taylor, Lloyd W., "The Untold Story of the Telephone." Am. J. Phys., 5:243-251, Dec. 1937. 10 illus., port. of Elisha Gray. Review of little-known experiments, apparatus, documents and circumstances concerning the early work of Bell and Gray, particularly the latter's telephone receivers. 26 footnote refs. to papers and patents.

1560 Aitken, William, Who Invented the Telephone? London, Glasgow: Blackie, 1939. 196 pp., illus. Alexander Graham Bell and Elisha Gray; inventions, events and controversies. Extracts from contemporary documents and technical literature, with refs. to inventions and claims.

1561 Dolbear, Amos E., The Telephone. With Directions for Making a Speaking Telephone. Boston: Lee and Shepard; New York: Charles T. Dillingham; London: Trübner, 1877. 128 pp., 17 illus. Early history of electricity and magnetism, theories of electricity, sound, resonance, tone composition, history of early telephones, the author's telephone, how to make a telephone.

1562 Marland, Edward A., "The Telephone." Early Electrical Communication, pp. 182-200. Events from 1875 to 1877, with mention of the induction balance. 55 footnotes. See 275.

1563 Prescott, George B., The Speaking Telephone, Talking Phonograph and Other Novelties. New York: D. Appleton, 1878. 431 pp., illus. Enlarged 2nd ed., The Speaking Telephone, Electric Light and Other Recent Inventions, 1879. 616 pp., illus.

1564 Gray, Elisha, Experimental Researches in Electro-Harmonic Telegraphy and Telephony, 1867-1878. New York, 1878. 96 pp., illus. About the author's work on the transmission of musical tones and other sounds, with a brief biography.

1565 Field, Kate, The History of Bell's Telephone. London: Bradbury, Agnew, 1878. 67 pp.

1566 Du Moncel, Theodore A. L., The Telephone, the Microphone and the Phonograph. London; New York: Harper and Brothers, 1879. 363 pp., 70 illus. First English edition, first work on the microphone and an early work on the phonograph.

1567 Thompson, Silvanus P., Philipp Reis, Inventor of the Telephone, a Biographical Sketch. London; New York: E. & F.N. Spon, 1883. 182 pp., 2 plates, illus. Support of the inventor's claim, with "documentary testimony, translations of the original papers of the inventor and contemporary publications." Refs. pp. 180-182.

1568 Prescott, George B., Bell's Electric Speaking Telephone: Its Invention, Construction, Application, Modification, and History. New York: D. Appleton, 1884. 526 pp., 330 illus.

1569 Benjamin, Park, "The Speaking Telephone." The Age of Electricity, pp. 272-324. The work of Reis, Gray, Bell, Dolbear, Hughes, and others, with mention of the phonograph and photophone. See 237.

1570 Preece, William H., and Julius Maier, The Telephone. London: Whittaker, 1889. 489 pp., illus.

1571 Walmsley, Robert M., "The Telephone." Electricity in the Service of Man, pp. 691-773. History, Bell's telephone and others, battery telephones, microphones, installations. Full description of instruments, apparatus, circuits and installations. 98 illus. See 238.

1572 Poole, Joseph, The Practical Telephone Handbook and Guide to the Telephonic Exchange. London: Whittaker, 1891. 288 pp., illus. Numerous eds.: 2nd, 1895, 347 pp.; 3rd, 1906, 533 pp.; 4th, 1910, 606 pp.; 5th, 1912, 624 pp.; 6th, 1919, 725 pp.; 7th, 1927, 870 pp. 8th ed. revised by N. V. Knight and W. Prickett. London: Pitman, 1942, 510 pp. A basic text.

1573 Blanchard, Julian, "A Pioneering Attempt at Multiplex Telephony." Proc. IEEE., 51:1706-1709, Dec. 1963. Story of several inventors' ideas during the 1890s, particularly those of John S. Stone. 9 footnotes.

1574 Rhodes, Frederick L., Beginnings of Telephony. New York; London: Harper and Brothers, 1929.

261 pp., 30 plates, 20 diags. App. A, pp. 207-224, list of important lawsuits with brief descriptions. App. B, pp. 225-233, early uses of the word "telephone." App. C, pp. 234-238, numerical list of U.S. patents, 86 entries from April 1875 to July 1918. App. D, pp. 239-244, sources of information consulted (refs.), 100 entries. A thorough study of progress in the United States up to the mid-1890s. The first 75 pages concern Bell and his patents. The remainder deal with microphones, overhead lines, cables, loaded lines, switchboards, station apparatus, the phantom circuit and long-distance lines. Much information on companies, people, patents and litigation; mechanical and electrical details, technical improvements, installations and operations; with numerous quotes and extracts from company reports, memoranda, addresses, technical journals, pamphlets, court records and other contemporary sources.

1575 Brittain, James E., "The Introduction of the Loading Coil: George A. Campbell and Michael I. Pupin." Technology and Culture, 11:36-57, Jan. 1970. 66 footnotes with extensive refs. and comments. Also, 11:596, 597; 601-603, Oct., 2 letters.

1576 Bell, Alexander G., The Bell Telephone. The Deposition of Alexander Graham Bell in the Suit Brought by the United States to Annul the Bell Patents. Boston: American Bell Telephone Company, 1908. 469 pp., illus.

1577 Casson, Herbert N., The History of the Telephone. Chicago: A. C. McClurg, 1910. 315 pp., 27 plates. A popular, nontechnical account without refs.

1578 Kempe, Harry R., and Emile Garcke, "The Telephone." Encyclopaedia Britannica, 1911. 11th ed., 26:547-557. 13 illus. Early history, transmitters, receivers, systems, circuits, British commercial aspects, companies. Brief chronology from 1876 to 1907. 15 footnote refs., 12 other refs.

1579 Webb, Herbert L., The Development of the Telephone in Europe. London: Electrical Press, 1911. 78 pp. A reprint of articles from Electrical Industries, 1910.

1580 Lawson, R., "A History of Automatic Telephony." J. POEE., 5:192-207, 1912-1913.

1581 Kingsbury, John E., The Telephone and Telephone

Exchanges: Their Invention and Development. London: Longmans, Green, 1915. 558 pp., illus.

1582 Kingsbury, John E., "The Story of the Telephone." Electrician, 77:812-814, Sept. 15, 1916.

1583 Fowle, Frank F., "Outline of the Most Striking Developments in Telephony during the Past Fifty Years." J. Western Society of Engineers, 24:497-499, Oct. 1919.

1584 Shaw, Thomas, "The Conquest of Distance by Wire Telephony." B. S. T. J., 23:337-421, Oct. 1944. "A story of transmission development from the early days of loading to the wide use of thermionic repeaters." 8 organization tables, map. Editorial foreword. 1904 to 1907 at Boston. The 1907-1911 period in New York. Transcontinental telephone project. Establishment of a transcontinental network. 3 Apps., report and memos. 11 footnotes. Bibliography, p. 407, 18 items.

1585 Lavine, A. Lincoln, Circuits of Victory. Garden City, N. Y.: Country Life Press, 1921. 634 pp., 32 plates, 11 illus. On the Bell Telephone System and telephone work during World War 1. Numerous extracts, no refs., no index.

1586 [Bell, A. G.]. "Bell and His Telephone; Little Known Facts." Sci. Am., 127:232, Oct. 1922. Brief sketch with 2 illus.

1587 Blake, George G., "The Development of the Telephone." History of Radio Telegraphy and Telephony, pp. 12-31. 17 illus. Experiments and apparatus from 1831 to 1922. Refs. to names, dates and patents. See 1193.

1588 Langdon, William C., "The American Telephone Historical Collection." Bell Tel. Qtly., Jan. 1924.

1589 Langdon, William C., "The Growth of the Historical Collection." Bell Tel. Qtly., April 1925.

1590 Baldwin, Francis G. C., The History of the Telephone in the United Kingdom. London: Chapman and Hall; New York: D. Van Nostrand, 1925. 728 pp., 75 plates, 94 diags., charts and maps, 24 tables and schedules. App. A, pp. 673-675, early telephone testimonials. App. B, pp. 675-682, draft of license issued to early telephone companies. App. C, pp. 682-690, draft of modified license. App. D, pp. 690-692, County of London by-laws on overhead wires, 1891. App. E, pp. 692-695, list of historic telephone exhibits at the Science Museum, December 1922, 75 items from 1818, with catalogue numbers. Also, list of historic telephone apparatus in the museum of the Institution of Electrical Engineers, 27 items; and 21 additional items presented to the Institution by the author. The first 120 pages deal with Bell's telephone and its introduction into England, 1876-1878, and early telephone companies in London and the provinces up to 1886. The next 70 pages cover developments up to 1890, including underground cables from 1823. The National Telephone Company, multiple switchboards, and dry core paper cables up to 1912 are treated in pp. 190-338. Pages 339-440 cover common battery systems, municipal telephones, the Post Office, and overhead line plant. The final parts are concerned with transmission, including loading coils, long-distance communication from 1879, distribution, automatic or machine switching from 1879, legislation and miscellaneous matters. Radio telephony is briefly mentioned (pp. 641-652), but almost wholly in regard to broadcasting. There are no formal references, but names, dates, patent numbers and other particulars from journals, books and contemporary sources, along with extracts, are plentiful in the text.

1591 Kingsbury, John E., "The Telephone, 1876-1926." Electrical Review, 98:364, 365, March 5, 1926.

1592 Harrison, H. H., "The Telephone, 1876-1926." Electrician, 97:4, 5, 14, July 2, 1926. Also, "Telephone Exhibition," pp. 6, 7, 14. 8 photos. Also, "Sir Oliver Lodge on Telephony," pp. 8, 9, 14.

1593 Lodge, Oliver J., "History and Development of the Telephone." J. IEE., 64:1098-1109, Nov. 1926. Also pp. 1093-1097.

1594 Jewett, Frank B., "The Telephone Switchboard--Fifty Years of History." Bell Tel. Qtly., 7:149-165, July 1928. 7 illus. Review from 1878; types and apparatus.

1595 Purves, Thomas F., "Telegraphy and Telephony. Steps in the Development of Wire-Borne Communications During the Past Century." Electrician, 107:414-417, Sept. 25, 1931. 10 photos. Brief outline of progress.

1596 [AT&T]. "The Telephone Industry." The Development of American Industries, pp. 721-743. Early history and development, the industry in 1927, economic significance, advances in long-distance transmission, the Bell System, world distribution of telephones (June 1931), research, radio telephony and services. Chronology from 1875. 1st ed., 1932, compare other eds. See 337.

1597 Langdon, William C., "Myths of Telephone History." Bell Tel. Qtly., 12:123-140, 1933.

1598 MacMeal, Harry B., The Story of Independent Telephony. Chicago: Independent Pioneer Telephone Association. 1934. 289 pp.

1599 Barrett, R. T., "Some Historic Telephone Exhibits." Bell Tel. Qtly., 18:147-160, July 1939. Covers over 50 years of history and various exhibitions.

1600 Jewett, Frank B., "A Quarter Century of Transcontinental Telephony." Elect. Eng., 59:3-11, Jan. 1940. 6 diags. A review of the significant developments in long-distance telephony since the first transcontinental telephone in 1914.

1601 Gray, G. H., "The Evolution of Wire Transmission." Elect. Comm., 20:235-245, 1942. Review of telephone transmission, carrier and coaxial cable, with comments on speech quality, system reliability, and economics. Bibliography.

1602 Angwin, A. Stanley, "Graham Bell--Pioneer: An Era of Outstanding Developments in World Communications." J. IEE., 94(Pt. 1):269-274, June 1947. Recapitulation of historical highlights and general discussion of telephone progress to 1943.

1603 Robertson, John H., The Story of the Telephone. A History of the Telecommunications Industry of Britain. London: Pitman, 1947. 299 pp. A general, nontechnical account by a political journalist who calls the story "a thin mock-up." No list of contents or chapter headings, no illus., no refs. Some names, dates, places, titles and extracts in the text.

1604 Andrews, E. G., "Telephone Switching and the Early Bell Laboratories' Computers." B. S. T. J., 42:341-

353, March 1963. 2 photos, table. Summary of 7 relay-type computers. Work and development from 1939 to 1950. 18 refs. from 1920.

1605 Martin, William H., "Seventy-five Years of the Telephone: An Evolution in Technology." B. S. T. J., 30:215-238, April 1951. 4 illus. Experiments, theory, design and measurements related to telephone instruments. Bibliography, 80 items from 1903.

1606 Wilkinson, Roger I., "Beginnings of Switching Theory in the United States." Elect. Eng., 75:796-802, Sept. 1956. 5 illus., 9 refs.

1607 Coulson, Thomas, "Franklin Institute Museum; Telephone Exhibit." J. F. I., 264:337, 338, Oct. 1957.

1608 [AT&T]. Events in Telephone History. New York: American Telephone and Telegraph Company, 1958, 1965. 20+52 pp. Alphabetical index, pp. iii-xx. Chronology of events (1790-1962), pp. 1-52, contains descriptive matter, in some instances with refs. Emphasis is upon company and public events, but some technical items are included.

1609 Romnes, Haakon I., "Advances in Communications." Elect. Eng., 78:481-492, May 1959. 19 illus. Progress in telephone technology over 25 years.

1610 Stecker, R. B., "A History of Transmission Systems." Western Electric Engineer, 3:3-10, Oct. 1959. Telephone history from 1876. 14 refs. from 1919.

1611 Hanscom, C. Dean, Dates in American Telephone Technology. New York: Bell Telephone Laboratories, 1961. 148 pp. Bibliography, pp. 146-148, 67 items, mostly Bell System or related company publications. A classified listing of products, events and inventions by name and date, with source references. A few non-Bell and foreign items are included.

1612 Abbott, Henry H., "Sixty Years of PBX Development." Bell Lab. Rec., 46:8-16, Jan. 1968. 11 photos. Bibliography of previous articles on PBX in the Record, p. 16, 27 entries from 1928.

1613 Carson, David N., "The Evolution of Picturephone

Service." Bell Lab. Rec., 46:282-291, Oct. 1968. 8 photos. From 1956, with mention of the picture transmission experiments of April 1927.

D. Telegraph and Telephone

1614 Tegg, William, Posts and Telegraphs, Past and Present; with an Account of the Telephone and Phonograph. London: W. Tegg, 1878. 318 pp.

1615 King, W. James, The Telegraph and the Telephone. Washington: Smithsonian Institution, 1962. United States National Museum Bulletin 228. Contributions from the Museum of History and Technology, Paper 29, pp. 273-332. 80 illus. A technical survey from the Chappe telegraph to the telegraph and telephone of the 1880s. Picture sources are identified, many are of items in the National Museum. 64 footnote refs. to books, papers and patents.

1616 Houston, Edwin J., The Electric Transmission of Intelligence. New York: W. J. Johnston, 1893. 330 pp., illus. Electric telegraph and telegraph cables, annunciators, alarms, time systems, the telephone, and other applications of electricity.

1617 Dibner, Bern, "Communications." Technology in Western Civilization, Vol. 1, pp. 452-460. This part of Chap. 27 briefly scans the highlights of the electric telegraph, telephone and the Atlantic cable. See 232.

1618 Beck, Arnold H. W., "The History of Telegraphy, The History of the Telephone." Words and Waves, pp. 45-73, 74-89. Early events and telephone history up to World War 1. See 310.

1619 Colpitts, Edwin H., and O. B. Blackwell, "Carrier Current Telephony and Telegraphy." Trans. AIEE., 40:205-300, Feb. 1921. Reprinted, Proc. IEEE., 52:340-359, April 1964. 25 illus. Historical treatment from Gray's multiple harmonic telegraph to commercial systems of the World War 1 period. Bibliography, 78 items in chronological order from 1867. 18 footnote refs. from 1886.

1620 Fleming, John A., "Telegraphs and Telephones from 1870 to 1920." Fifty Years of Electricity, pp. 50-108. Telegraph apparatus; transmitters, receivers, relays, printers, recorders. Submarine cables, cable ships and the cable industry. Telegraph systems; Wheatstone, Edison, Baudot, Murray. Telephone inventions and apparatus, receivers, transmitters, sets, lines, switchboards. Systems, operations, circuits. Loading coils and long-distance lines, automatic switching systems. Contributions of Bell, Edison, Hughes, Heaviside, Pupin. 22 plates and 39 diags. showing equipment, installations, instruments, cables, waveforms, codes and circuits. See 261.

1621 Denman, R. P. G., The History of Electrical Communication. Catalogue of the Collections in the Science Museum, South Kensington; with Descriptive and Historical Notes and Illustrations: Line Telegraphy and Telephony. London: H. M. S. O., 1926. 55 pp.

1622 Sellars, H. G., "A Brief Chronology for Students of Telegraphs, Telephones and Posts." Telegraph and Telephone J. A serial listing from 14:43, 44, 1927.

1623 Jewett, Frank B., "Telegraphy and Telephony." A Century of Industrial Progress, pp. 445-463. A retrospective view of American developments, including radio. See 336.

1624 Michaelis, Anthony R., "The Telegraph and the Telephone, 1793-1932." From Semaphore to Satellite, pp. 11-116. Early history, the pioneers, beginnings of international co-operation, the Paris conference 1865, the Telegraph Union, telegraph rates, the telephone and regulations. 135 photos, maps, engrs., facsimiles. See 308.

16. TELEVISION AND FACSIMILE

Television, like radio, has been well treated in general literature since the mid-1920s. A few popular articles, selected from hundreds of the period, are included in this chapter to represent contemporary reports on individual systems and demonstrations, roughly from 1923 to 1935. Biographies of note include the following (Sect. 4C): Baird, 381-384; Farnsworth, 404; Ives, 422; Jenkins, 423; Korn, 425; Swinton, 451; Zworykin, 464. See also 530 (Cossor), 531 (Du Mont), 550-552 (RCA), for company activities. Contemporary practices are covered in several handbooks (Sect. 1E): 48, 49, 56-58, 64; and in general histories (Chap. 3): 283, 295, 323, and others. For broadcasting history (Chap. 6), see 587, 589, 593, 618, 619, and most of the entries in Sect. 6C. Related material is contained in 755, 799, 804, 806, 1193, 1212, 1216, 1417, 1420. Extracts from some of these items are listed in this chapter.

A. General Surveys

Numerous books and articles published between 1926 and 1936 contain historical surveys. Those that pertain solely to mechanical television or specific systems, as well as facsimile, are entered in Sect. 16B.

1625 Moseley, Sydney A., and H. J. Barton Chapple, Television To-day and To-morrow. London: Pitman, 1930. 130 pp., 38 diags., 47 plates. Almost wholly about Baird and his system. 4th ed., 1934. 208 pp., 56 diags., 71 plates. 5th ed., 1940. 179 pp., 56 diags., 34 plates, folding map of television service area, Alexandra Palace. History, principles, scanners and cameras, antennas, cathode-ray tubes, receivers, big-screen television, special television methods. App. pp. 167-175, British Standards, receiver control markings, standard television terms. Popular style, intended for the general reader and amateur constructor. Each ed. includes history.

1626 Dinsdale, Alfred, "Television in America To-day." J. Television Soc., 1:137-149, 1932. 15 photos and diags. App. 1, List of experimental stations, 24 entries with call letters, power and frequency. App. 2, CBS program for Tuesday, Dec. 29, 1931. Two-way television over telephone circuit (BTL), Farnsworth system, Sanabria projection system, disk receivers, future American plans for cathode-ray television, commercial development and government control.

1627 "The Chronology of Television." Television and Short-Wave World. 8:391-393; 453,454, July, Aug. 1935. 3 illus. 115 entries by year from 1814.

1628 Washburne, R.D., and Wilhelm E. Schrage, "World-Wide Television." Radio-Craft, 7:76-80,123-126, Aug. 1935. 44 illus. "Television is gaining international momentum." Description of the work and progress in 12 countries. List of 25 television stations in the U. S.

1629 Gorokhov, P.K. (O.M. Blunn, trans.), "The Origins of Modern Television." Radio Engineering, 16:104-116, June 1961 (Radiotekhnika, 16:70-79, 1961). Port., Boris Rosing. Early history with emphasis on Rosing's work, development of mechanical systems, and Russian contributions up to 1935. 27 refs., including patents, from 1873.

1630 [RCA]. Television. Princeton, N.J.: RCA Review. 6 vols., each containing reprints or summaries of technical articles by RCA personnel. Vol. 1, 1933-1936; 2, 1937; 3, 1938-1941; 4, 1942-1946; 5, 1947-1948; 6, 1949-1950. Several papers are listed separately in this chapter.

1631 Clarke, Basil, "The First Live Transatlantic Link." International TV Technical Rev., 3:498-501, Oct. 1962. 3 photos. Also "The Quest for High Definition," pp. 546-549, Nov. 1962. 2 photos. Background of Baird's work, transatlantic tests, early British broadcasts and events to 1936.

1632 Ashbridge, Noel, "Television in Great Britain." Proc. IRE., 25:697-707, June 1937. Survey from the first Baird demonstrations with the B. B. C. in 1929.

1633 Garratt, Gerald R.M. (ed.), and Geoffrey Parr, Television. London: H.M.S.O., 1937. 64 pp., 16 illus. "An account of the development and general principles of television as illustrated by a special exhibition held at the Science Museum, June-September, 1937." Chap. bibliographies contain 65 items, some referenced in the text. A short survey from the 1880s up to the London television station opening in 1936. See next entry.

1634 [Science Museum]. "Television Exhibition." J. Television Soc., 2:265-273, 1937. A report by the editor on the exhibition, June to Sept. 6 photos. Detailed descriptions of 15 exhibitors' products; Baird, Edison, Swan, EMI, General Electric, RCA, Scophony, Standard Telephones and Cables, including historical items.

1635 Wilson, John C., Television Engineering. London: Pitman, 1937. 492 pp., 276 photos, graphs, diags. A comprehensive text devoted almost wholly to mechanical systems, with full analyses of principles and techniques covering a wide range of proposals, inventions and methods. Chap. 1, pp. 1-12, Historical. 37 refs. from 1845, including patents and articles. Chap. 2, pp. 13-45, The eye and optics. Chap. 3, pp. 46-72, Scanning methods and devices. Chap. 4, pp. 73-108, Analysis of finite aperture scanning methods. Chap. 5, pp. 109-155, Photosensitivity. Chap. 6, pp. 156-221, Amplifiers, channels and filters. Chap. 7, pp. 222-257, Light modulation. Chap. 8, pp. 258-316, Cathode rays and fluorescence. Chap. 9, pp. 317-354, Synchronizing. Chap. 10, pp. 355-388, Special television methods. Chap. 11, pp. 389-419, Modern television equipment. Chap. 12, pp. 420-437, Physical limitations. Apps., pp. 438-463, 9 items. App. 7, pp. 460, 461, Table of early television disclosures, 40 entries from 1877, with refs. Two folded charts on scanning systems contain 363 entries, mostly patent numbers, in 58 categories. 593 chap. refs. See also 1725.

1636 "Mileposts in Television." Radio-Craft, 9:576-579, 607, March 1938. 38 small illus. Survey of progress from Nipkow's proposal to current electronic systems.

1637 Macnamara, T. C., and D. C. Birkinshaw, "The London Television Service." J. IEE., 83:729-757, Dec. 1938. Description of the B. B. C. television station at Alexandra Palace, North London, with a survey of progress from the inception of low-definition television by Baird's mechanical system (1929).

1638 Archer, Gleason L., "Historical Background of Television." Big Business and Radio, Chap. 21, pp. 430-446. Early history, mechanical scanning, RCA and electronic television. Also, Chap. 22, pp. 448-466, Television and facsimile. Iconoscope, RCA receivers, Farnsworth system, television broadcasting, coaxial cable, radio relays, radio facsimile and recording methods. List of TV stations, Feb. 1, 1939. See 1417.

1639 Briggs, Asa, "The New World of Television." The History of Broadcasting in the United Kingdom, Vol. 2, pp. 519-622. Detailed treatment of men and events from 1925 to 1939. Baird and his associates, B. B. C. personnel, Marconi-EMI, technical developments, experiments, agreements, correspondence, reports, news announcements, programs. Copious footnotes and source refs. See 582.

1640 "Television I: A $13,000,000 'If'." Fortune, 19:52-59, 168, 172-182, April 1939. 19 illus. A preview of regular U. S. television service, its problems and possibilities, with some historical background on inventors and companies. The second part published in May deals with studio and entertainment aspects of the new medium.

1641 Town, George R., "Television." Elect. Eng., 59: 313-322, Aug. 1940. 10 illus., 3 tables. A quick look at history from the 1870s and a brief description of operating principles followed by a discussion of contemporary practices in the United States with mention of British television. No refs.

1642 Engstrom, Elmer W., "Recent Developments in Television." Ann. Am. Acad., 213:130-137, Jan. 1941. A general nontechnical survey of current television. See 584.

1643 Hylander, Clarence J., and Robert Harding, An Introduction to Television. New York: Macmillan, 1941. 207 pp., 79 illus. Popular, semitechnical treatment with some history (The birth of television, pp. 21-55). No refs.

1644 de Forest, Lee, Television Today and Tomorrow. New York: Dial Press, 1942. 361 pp., 77 illus. A general outline of systems and techniques with a little history. A semitechnical exposition for the layman. Not documented.

1645 Dunlap, Orrin E., The Future of Television. New York: Harper and Brothers, 1942. 194 pp., 8 plates. App., pp. 180-186, Historic steps in television, 100 entries by year from 1867. List of commercial television stations (U. S.), p. 187, 11 entries. A popular survey with some history. Some dates given as footnotes, extracts in the text, otherwise no refs.

1646 Hubbell, Richard W., 4000 Years of Television. The Story of Seeing at a Distance. New York: G. P. Putnam's Sons, 1942. 256 pp., 8 diags. Popularly written, with nontechnical explanations and a typical account of the history of electricity and communications. The American story is supplemented by chapters on foreign television, pp. 159-189; and television in wartime, pp. 190-212. No refs.

1647 Rettenmeyer, Francis X., "Radio-Electronic Bibliography." Radio, pp. 43, 44, 46, 48, 50; 50, 52, 54, 56; 44, 46, 48, 50; 54, 56, 58, Aug. to Nov. 1943. 12, Television, 1231 items from 1922. See 93.

1648 Official Yearbook of the Television Industry. New York: Television Broadcasters Association. Published from 1945. 1944-1945, 75 pp. Short articles on television topics, with a chronology of events related to television from 1867.

1649 Engstrom, Elmer W., "Television--A Review, 1946." In Television, Vol. 4 (1942-1946), pp. 467-481. 13 illus. A brief review of RCA developments from 1939, with details of progress in camera tubes, transmitters, receivers and network facilities. No refs. See 1630.

1650 American Television Directory. New York: American Television Society, 1946. Official Yearbook, first ed., 144 pp. General business information and numerous articles on a variety of television topics.

1651 Zworykin, Vladimir K., "Television--Retrospect and Prospect." Tech. Rev., 49:333-336, 354, April 1947. 2 illus. Historical survey with a nontechnical account of camera tube and circuit developments, prospects for color television and television broadcasting. 4 footnotes.

1652 Town, George R., "Progress in Television." Elect. Eng., 66:580-590, June 1947. 10 diags., 2 tables. Television standards, receiver manufacture, wartime improvements, networks and programming, color television and systems. 5 refs.

1653 Kempner, Stanley, "Milestones to Present-Day Television." Television Encyclopedia, Pt. 1, pp. 3-42. Chronological survey from 600 B. C. to 1947. See 64.

1654 Lankes, L. R., "Historical Sketch of Television's Progress." J. SMPE., 51:223-229, Sept. 1948. Brief outline of highlights, including facsimile, from 1842. 22 refs. from 1897. In 1671.

1655 Parr, Geoffrey, "British Television." Discovery, 10(NS):212-218, July 1949. 12 photos and diags. History, technical details, equipment. Reading list, 6 items from 1839.

1656 Maclaurin, William R., "The Rise of Industrial Research--Television." Invention and Innovation in the Radio Industry, pp. 191-224. 6 diags. 5 tables of statistics; RCA, Farnsworth, CBS, Du Mont. Survey of European origins and progress by U. S. companies. Also "Government Regulation and Technical Progress--FM and Television: 1900-1941," pp. 225-240. See 1212.

1657 [I. E. E.]. "British Contribution to Television." Proc. IEE., 99(Pt. 3A):1-866, May 1952. Proceedings at the IEE convention, London, April 28 to May 2. 85 papers. See next entry.

1658 Garratt, Gerald R. M., and A. H. Mumford, "The History of Television." Proc. IEE., 99(Pt. 3A):25-42, May 1952. 16 illus. Broad review of early facsimile, discovery of the effect of light on selenium, early mechanical systems, low-definition systems of the mid-1920s, electronic systems, and public television service in Britain. Topics include proposals by Senlecq, Ayrton and Perry, Carey, Bidwell, Nipkow, Weiller, Rosing, Swinton; systems of Mihaly, Jenkins, Baird, Ives (B. T. L.); high-definition systems of Bedford and Puckle (Cossor), Baird, E. M. I., Zworykin (RCA), Scophony. Several extracts, including part of the Selsdon Report (1935). 45 refs. from 1839. Discussion (pp. 40-42), 5 footnotes. A basic paper.

1659 Jensen, Axel G., "The Evolution of Modern Television." J. SMPTE., 63:174-188, Nov. 1954. 29 illus. Full survey of electronic systems, with emphasis on American work. Brief mention of early mechanical proposals. Other topics are Swinton's proposal, early work of Zworykin, von Ardenne, Farnsworth, flying-spot scanner, the Davisson tube, dissector tube, iconoscope, image iconoscope, orthicon, image orthicon, vidicon, multiplier phototubes, motion-picture film scanners, television standards and network facilities in the U. S. Various extracts. 40 refs. from 1906. A basic paper. In 1671.

1660 Abramson, Albert, "A Short History of Television Recording." J. SMPTE., 64:72-76, Feb. 1955. Historical development from 1927, early film recording in the U. S. and Britain; modern magnetic video recording. 27 refs. from 1927. In 1671.

1661 Abramson, Albert, Electronic Motion Pictures: A History of the Television Camera. Berkeley, Los Angeles: University of California Press, 1955. 212 pp., 93 illus. A thorough survey of proposals, systems, experiments, innovations and practices. Includes studio techniques, recording methods, large-screen television, airborne television, commercial operations from 1946 in the U. S., England and Germany, and details of processes and equipment of the early 1950s. Chap. 2, pp. 15-24, Progress to 1900. Chap. 3, pp. 25-34, Experimentation and theory (1900-1923). Chap. 4, pp. 35-50, Television becomes a reality (1923-1929). Chap. 5, pp. 51-64, Early television systems and film. Chap. 6, pp. 65-85, The introduction of the electronic camera. Chaps. 7 to 10 cover progress from 1937; improved cameras, film recording, large-screen television, commercial operations to 1955. Chap. notes, pp. 179-198, contain 375 refs. to books, articles and patents, some multiple entries. A basic book.

1662 Pioneering in Television. Prophecy and Fulfillment. New York: RCA, various years. 4th ed., 1956, 177 pp., 75 illus. Nearly 100 excerpts from speeches and statements by Brigadier General David Sarnoff. List of RCA-NBC firsts in television, pp. 144-154, 101 entries from December 1923. List of a few historical documents, pp. 155-176, 7 items.

1663 Shunaman, Fred, "30 Years of Television." Radio-Electronics, 28:50-53, Jan. 1957. 7 illus.

1664 [Bell Laboratories]. "Television: 30 Years of Progress." Bell Lab. Rec., 35:150-152, April 1957. 5 photos. Highlights of Bell contributions from 1927.

1665 Sturmey, S. G., "Broadcasting: Television." The

Economic Development of Radio, Chap. 10, pp. 190-213. A little early history from 1873, Baird Company, B. B. C., Marconi-E. M. I., Cossor, Scophony, television committees, public programing. 28 refs. See 1420.

1666 Zworykin, Vladimir K., Edward G. Ramberg, and Leslie E. Flory, Television in Science and Industry. New York: John Wiley, 1958. 300 pp., 206 illus. History, pp. 1-21, from the mid-19th century, is a brief survey with 19 refs. and 5 footnotes. Over 200 chap. refs.

1667 Pawley, Edward L. E., "B. B. C. Television 1939-60; a Review of Progress." Proc. IEE., 108(Pt. B):375-397, July 1961. 2 maps, 2 charts, table of B. B. C. television stations. List of outstanding dates in television, 31 entries by full date from Sept. 1, 1939. 49 refs. from 1938. 7 footnotes, patents. See also 1669.

1668 Bingley, Frank J., "A Half Century of Television Reception." Proc. IRE., 50:799-805, May 1962. Early mechanical systems, development of electronic television, television receivers and circuits, television standards and color television up to 1960. 78 refs. from 1927.

1669 BBC Television. A British Engineering Achievement. London: B. B. C., 1961. 64 pp., 55 photos, diags. and maps, some in color. The early days up to 1939 and the post-war years. Studios, regional centers, equipment for special effects, telerecording, film making, outside broadcasts, television service, international exchange of programs. Chronology, pp. 61-64, Some important dates in the development of BBC television, 102 dated entries from Nov. 2, 1936. See also 1667.

1670 Bitting, Robert C., "Creating an Industry." J. SMPTE., 74:1015-1023, Nov. 1965. Pt. 1, Formation and growth of RCA. Pt. 2, Television development within RCA. Mechanical television, early electronic television, field tests 1933-1939, inauguration of a public service, the war years, the postwar era, color television. 21 refs. from 1929.

1671 Fielding, Raymond (comp.), A Technological History of Motion Pictures and Television. An anthology from the pages of the Journal of the Society of Motion Picture and Television Engineers. Berkeley and Los Angeles: University of California Press, 1967. Pt. 3, pp. 227-254,

Television, comprises 4 papers listed separately (1654, 1659, 1660, 1678). Bibliography of additional papers, p. 255, 48 items.

B. Mechanical Television

A variety of proposals for transmitting images by electricity were made public from 1877. Most of these were intended as facsimile systems whereby graphic images were to be reproduced (Sect. 16D). Successful demonstrations of the transmission of moving images in halftone were accomplished in 1926. Television thereafter progressed from crude experiments to public transmissions with highly engineered equipment by the mid-1930s. History of this mechanical era is included in several items listed in Sect. 16A, especially 1625, 1638, 1639, 1658, 1659, 1661, 1668. Almost all of 1635, 1647, and 1653 are concerned with this period. Electromechanical and semi-electronic systems are included in this section. See also 1799.

1672 Shiers, George, "Early Schemes for Television." IEEE Spectrum, 7:24-34, May 1970. 11 illus. Early proposals from 1877 to 1884. Chronological list, 18 items. 51 refs. from 1840. Also, Proc. IREE Australia, 31:407-416, Dec. 1970. On the work of Ayrton and Perry, Bidwell, Carey, Lucas, Nipkow, Redmond, Sawyer, Senlecq.

1673 Le Pontois, Leon, "The Telectroscope." Sci. Am. Supp., 35:546, 547, June 10, 1893. 3 diags. showing pin-hole disks, transmitter and receiver.

1674 "The Dussaud Teleoscope." Sci. Am. Supp., 46: 18793, July 2, 1898. 2 illus. of apparatus.

1675 [Szczepanik, Jan]. "The Telectroscope and the Problem of Electrical Vision." Sci. Am. Supp., 46: 18889, 18890, July 30, 1898. 10 illus. showing constructional details.

1676 [Senlecq, Constantin]. "The Senlecq Telectroscope. An Apparatus for Electrical Vision." Sci. Am. Supp., 64:372, 373, Dec. 14, 1907. 6 illus. of assembly.

1677 [Rosing, Boris]. Grimshaw, Robert, "The Telegraphic Eye." Sci. Am., 104:335,336, April 1, 1911. 2 views of Rosing's apparatus and a circuit diag. Also, Sci. Am. Supp., June 17, 1911. Also, Ruhmer, Ernest, "An Important Step in the Problem of Television." Sci. Am., 105: 574, Dec. 23, 1911. Photo of Rosing's mirror-drum transmitter and one of his cathode-ray tube receiver.

1678 Hogan, John V. L., "The Early Days of Television." J. SMPTE., 63:169-173, Nov. 1954. 3 illus. On the mosaic and scanning disk systems, with a brief survey of the methods tried out during the 1920s; Carey, Bidwell, Nipkow, Jenkins, Baird, Bell Telephone Laboratories, General Electric, RCA. No refs. In 1671.

1679 Davis, Watson, "The New Radio Movies." Popular Radio, 4:436-443, Dec. 1923. 8 illus. Laboratory demonstration of "shadowy motion" by means of the Jenkins multiple-lens disk. Also, Gernsback, Hugo, "Radio Vision." Radio News, 5:681,823, Dec. 1923. 5 illus.

1680 Baird, John L., "An Account of Some Experiments in Television." Wireless World, 14:153, May 7, 1924.

1681 Langer, Nicholas, "Radio Television." Radio News, 5:1570,1571,1686-1689, May 1924. 9 photos and diags. On the experiments of Dionys Mihaly and his Telehor apparatus.

1682 O'Connor, Sexton, "Experiments in Television." Popular Wireless and Wireless Rev., May 31, p.504; June 7, pp.539,540,558, 1924. 3 photos, 5 diags. Survey of various systems: Belin, Korn, Jenkins, Rosing, Swinton, Mihaly, Baird.

1683 Baird, John L., "Television, or Seeing by Wireless." Discovery, 6:142,143, April 1925. Diag. of Baird's system, photo of apparatus. Brief announcement by the inventor, with observations by the editor. Also, "Wireless Television; Baird Apparatus." Engineering, 119:661,662, May 29, 1925. Illus. See also, Wireless World, 15:533, Jan. 21, 1925.

1684 Bird, P.R., "Wireless Television: A Review of the Baird System." Popular Wireless and Wireless Rev., May 23, pp.622,623, 1925. 6 photos. Report on interview and demonstration at Selfridges, London.

1685 Dinsdale, Alfred, Television. Seeing by Wire or Wireless. London: Pitman, 1926. 62 pp., 12 plates. In this first book on television the author is almost wholly concerned with Baird's system. 2nd ed., Television Press, 1928. 180 pp., plates, diags.

1686 Gradenwitz, Alfred, "The Mihaly Television Scheme." Popular Wireless and Wireless Rev., June 26, pp. 617-619, 1926. Port., 2 photos of equipment, one of reproduced letters.

1687 Dinsdale, Alfred, "Television an Accomplished Fact." Radio News, 8:206, 207, 280, 282, 283, Sept. 1926. 6 illus. Description of Baird's apparatus and demonstration in London.

1688 Fournier, L., "New Television Apparatus." Radio News, 8:626, 627, 739, 740, Dec. 1926. 6 illus. Latest developments of Belin and Holweck.

1689 Baird, John L., "Television." J. Sci. Insts., 4:138-143, Feb. 1927. 6 diags. Historical survey and brief description of Baird's apparatus.

1690 Alexanderson, Ernst F. W., "Radio Photography and Television." Radio, 9:18, 19, 48, 50-52, Feb. 1927. 4 photos. General discussion, with brief mention of the author's system (G. E.).

1691 Dinsdale, Alfred, "Television Sees in Darkness and Records Its Impressions." Radio News, 8:1422, 1423, 1490-1492, June 1927. 5 illus. Baird's Noctovision, or infrared pickup, and his Phonoscope, or phonograph recorder. Also, Sci. Am., 83:282, April 1927.

1692 Secor, H. Winfield, "Radio Vision Demonstrated in America." Radio News, 8:1424-1426, 1480, June 1927. 10 photos and diags. Particulars of the BTL system and demonstration in New York, April 7.

1693 Secor, H. Winfield and J. H. Kraus, All About Television. New York: Experimenter Publishing Co., 1927. 112 pp., many photos and diags. A survey of phototelegraphy and television, largely compiled from articles in Radio News, Science and Invention, and other Gernsback magazines. Television experiments, pp. 7-11. Baird television, experimental data, pp. 11-13, Television, pp. 60-104.

Other parts concerned with facsimile methods are listed in Sect. 16D (1802).

1694 Tiltman, Ronald F., Television for the Home. The Wonders of "Seeing by Wireless." London: Hutchinson, (1927). 106 pp., 8 plates. Primarily about Baird's work. Numerous extracts from magazines and newspapers, with passing references in the text. No index.

1695 Ives, Herbert E., "Television." B. S. T. J., 6:551-559, Oct. 1927. General survey of the problem and introduction to 4 papers. 9 footnotes. Gray, Frank, J. Warren Horton, and R. C. Mathes, "The Production and Utilization of Television Signals," pp. 560-603. 30 photos, graphs, diags. Description of apparatus, signal waveform, terminal circuits. Stoller, H. M., and E. R. Morton, "Synchronization of Television," pp. 604-615. 10 illus. General requirements, motor design, framing of picture, receiver and amplifier circuits, operation. Gannett, Danforth K., and Estill I. Green, "Wire Transmission System for Television," pp. 616-632. 10 illus., including map of circuits between Washington, New York, and Whippany, N. J. Wire lines, general requirements, circuit details, measurements. Nelson, Edward L., "Radio Transmission System for Television," pp. 633-652. 9 photos and diags. Apparatus, tests, circuits, transmission characteristics.

1696 Jenkins, C. Francis, "Radio Vision." Proc. IRE., 15:958-964, Nov. 1927. On the author's system. 6 illus. No refs.

1697 "Television 1873-1927." Television, 1:10, 11, 23, March 1928. A brief outline of mechanical television and photoelectric cells. 2 illus.

1698 Dinsdale, Alfred, "Seeing Across the Atlantic Ocean!" Radio News, 9:1232, 1233, May 1928. 2 photos, diag. Description of experimental transmission from Coulsdon, Surrey, to Hartsdale, N. Y., Feb. 8, and from the liner Berengaria, March 7.

1699 Larner, E. T., Practical Television. London: Ernest Benn, 1928. 175 pp., 97 illus., including 13 plates. App., p. 172, On the transatlantic test (Feb. 9, 1928). Chap. 2, pp. 35-62, Historical, primarily on facsimile methods. Other parts also give history: Chap. 5, pp. 91-108, Continental and American researches; Chap. 6, pp. 109-118, Researches with the cathode-rays. An excellent survey of facsimile and mechanical television up to 1927.

1700 [Jenkins, C. Francis]. "Radio Movies and Television for the Home." Radio News, 10:116-118, 173, Aug. 1928. 4 photos, 2 diags. Description of the Jenkins system.

1701 [Sanabria, Ulysses A.]. "Successful Television Accomplished on Broadcast Band." Radio News, 10:219, 220, 277, Sept. 1928. 4 photos. Particulars of apparatus and demonstration in Chicago.

1702 Tiltman, Ronald F., "Television in Natural Colors Demonstrated." Radio News, 10:320, 374, Oct. 1928. 2 illus. Description of Baird's apparatus and demonstration in London, June 11.

1703 [Westinghouse Electric and Manufacturing Co.]. "Radio 'Movies' from KDKA." Radio News, 10:416, 417, Nov. 1928. 3 photos. Demonstration of Westinghouse system in East Pittsburgh, Aug. 8.

1704 Tiltman, Ronald F., "How 'Stereoscopic' Television is Shown." Radio News, 10:418, 419, Nov. 1928. 2 photos of transmitter and receiver, diag. of transmitter. Description of Baird's apparatus and demonstration in London, Aug. 10.

1705 Hertzberg, Robert, "Successful Television Programs Broadcast by Radio News Station WRNY." Radio News, 10:412-415, 490-492, Nov. 1928. 5 photos, 2 facsimile extracts from the New York Times, Aug. 13, 21. Particulars of Pilot apparatus, demonstrations, programing, printed schedules.

1706 Hertzberg, Robert, "Television Makes the Radio Drama Possible." Radio News, 10:524-527, 587, 588, Dec. 1928. 6 photos, 2 diags. Details of GE (Alexanderson) equipment and stage play "The Queen's Messenger," broadcast from WGY, Schenectady, Sept. 11.

1707 Ives, Herbert E., "Television in Colors." Bell Lab. Rec., 7:439-444, July 1929. 8 illus. Brief description of equipment and process. Also, Telegraph and Telephone Age, No. 14:315-318, July 16, 1929. 3 illus. Telephone Engineer, 33:19, 20, 44, Aug. 1929. 4 illus.

1708 Baird, John L., "Television Enters Public Life." Discovery, 10:362, Nov. 1929. 3 photos. Editorial comment, pp. 362-364, with historical matter.

1709 Jenkins, C. Francis, Radiomovies, Radiovision, Television. Washington: National Capitol Press, 1929. 143 pp., port., numerous photos and diags. Details of the author's equipment, experiments and demonstrations.

1710 Zworykin, Vladimir K., "Television with Cathode-Ray Tube for Receiver." Radio Engineering, 9:38-41, Dec. 1929. Also, "Cathode-Ray Television Receiver Developed," Sci. Am., 142:147, Feb. 1930. 2 photos. Condensed report on the Westinghouse system. "Television Through a Crystal Globe." Radio News, 11:905, 949, 954, April 1930. 4 illus. Radio-Craft, 1:384, Feb. 1930.

1711 Yates, Raymond F., ABC of Television or Seeing by Radio. New York: Norman W. Henley; London: Chapman and Hall, 1929. 210 pp., 78 diags., 13 plates. Telephotography and mechanical television. Theory, construction and operation, "written especially for home experimenters, radio fans and students." A popular exposition with light technical treatment. No refs.

1712 Sheldon, Harold H., and Edgar N. Grisewood, Television, Present Methods of Picture Transmission. New York: D. Van Nostrand, 1929. 194 pp., 129 illus. Historical background, pp. 5-18, primarily facsimile. Baird system, pp. 118-131; Bell system, pp. 132-158; Jenkins system, pp. 159-165; Alexanderson system, pp. 166-174.

1713 Ives, Herbert E., and others, "Two-Way Television." Trans. AIEE., 49:1563-1576, Oct. 1930. Pt. 1, Image transmission system. Pt. 2, Synchronization system. Pt. 3, Sound transmission system. 21 illus. Also, Bell Lab. Rec., 8:399-404, May 1930. 5 illus. "Two-Way Television." Sci. Mon., 30:476-480, May 1930. 5 photos, diag. Report on the telephone-television system demonstrated in April.

1714 "Practical Television is Now Here!" Radio News, 12:26-29, July 1930. 21 photos. Pictorial essay on current television apparatus.

1715 Barnard, George P., "Use of Selenium Cells for Television." The Selenium Cell, pp. 267-288. 23 illus.

Description of various systems and circuits, with history. 40 refs. from 1879, pp. 290, 291. See 755.

1716 Du Mont, Allen B., "Practical Operation of a Complete Television System." Radio Engineering, 11:33-36, July 1931. 8 illus. Description of current Jenkins system.

1717 Ardenne, Manfred von, "The Cathode Ray Tube Method of Television." J. Television Soc., 1:71-74, 1931. 5 illus. Report on contemporary experiments by this pioneer, with brief mention of earlier work since 1928. 2 refs.

1718 Felix, Edgar H., Television. Its Methods and Uses. New York: McGraw-Hill, 1931. 272 pp., 73 illus. A general account of contemporary mechanical television systems with some history, also a view of television problems, requirements, applications and future progress.

1719 Wenstrom, William H., "The March of Television." Radio News, 13:752, 753, 810; 852, 853, 876-878, March, April 1932. 12 illus. A brief survey from Bain's printing telegraph through Bidwell's and Nipkow's experiments to Swinton's proposal and current work of Baird, Jenkins, General Electric and Bell Telephone Laboratories.

1720 [Baird, John L.]. "The Derby by Television." Discovery, 13:233-235, July 1932. 2 photos, diag. of system. Description of transmission from Epsom Downs to the Metropole Cinema, London, June 1.

1721 Chapple, H. J. Barton, "Recent Advances in Television." J. Television Soc., 1:106-112, 1932. 4 photos. History from early 1929, growth of broadcast service, daylight television, Derby transmission, large-screen television.

1722 Dunlap, Orrin E., The Outlook for Television. New York; London: Harper and Brothers, 1932. 297 pp., plates, ports. Layman's explanation of television. Developments from 1928 through 1931 are covered in several chapters. Includes a chronology of radio and television from 640 B. C.

1723 Zworykin, Vladimir K., "Description of an Experimental Television System and the Kinescope." Proc. IRE., 21:1655-1673, Dec. 1933. 18 photos, graphs, diags. Scanning disk and photocell transmitter with a cathode-ray tube receiver. 6 footnote refs. Also, Kell, R.D., "Description of Experimental Television Transmitting Apparatus," pp. 1674-1691. 18 illus. Beers, G.L., "Description of Experimental Television Receivers," pp. 1692-1706. 16 illus.

1724 Reyner, John H., Television, Theory and Practice. London: Chapman and Hall, 1934. 196 pp., plates, diags. 2nd ed., 1937. 224 pp., plates, ports., diags. An early text with a practical approach.

1725 Wilson, John C., "Twenty Five Year's Change in Television." J. Television Soc., 2:86-93, 1935. 3 illus. 37 refs., including numerous patents, from 1845. Table of early disclosures has 57 entries in 5 categories from 1839. Same as Chap. 1 and App. 7 in Television Engineering. See 1635.

1726 Lee, H.W., "The Scophony Television Receiver." Nature, 142:59-62, July 9, 1938. 2 diags. Technical outline with details of principles. 4 footnotes (patents). Also, Proc. IRE., 27:483-500, Aug. 1939. 4 articles, 32 illus.

1727 Ives, Herbert E., "Television: Twentieth Anniversary." Bell Lab. Rec., 25:190-193, May 1947. Review of early developments from 1925, especially the tests of 1927. See also, Ives, Ronald L., "Television's 40th Birthday." Radio-Electronics, 38:55, 56, April 1967. 4 photos. See 1695.

C. Electronic Television

This section is devoted solely to all-electronic systems. These became practicable with the invention of camera tubes during the 1920s and early 1930s; papers concerned with these tubes are therefore included. Items on picture tubes, employed with mechanical transmitting equipment, are listed in Sect. 16B. Other entries on electronic television are in Sect. 16A; see 1625, 1630, 1632-1635, 1637, 1639-1642, 1649, 1656-1659, 1661, 1662, 1665-1669. See also 770, 773, 774, 782, 806.

1728 Swinton, A.A. Campbell, "Presidential Address." J. Röntgen Soc., 8:1-15, Jan. 1912. This historic document has a schematic diagram of the electric television system previously suggested by the author. See "Distant Electric Vision," Nature, 78:151, June 18, 1908.

1729 Swinton, A. A. Campbell, "The Possibilities of Television." Wireless World, 14:51-56; 82-84, April 9, 16, 1924. Discussion, 114-118, April 23, 1924. This updated version of the author's earlier scheme (previous entry) has 3 diagrams showing electronic television circuits for line and radio transmission. See also, Model Engineer, 50:671, June 12, 1924.

1730 Swinton, A. A. Campbell, "Television: Past and Future." Discovery, 9:337-339, Nov. 1928. A quick look at history, some adverse comments on mechanical methods, and mention of the 1908 proposal.

1731 [Farnsworth, Philo T.]. "New Television System." Radio News, 10:637, Jan. 1929. Brief reference to press dispatches and 2 photos of Farnsworth and his equipment.

1732 Farnsworth, Philo T., and Harry R. Lubcke, "Transmission of Television Images." Radio, 11:36, 85, 86, Dec. 1929. 2 illus., one of early electronic equipment, the other a photo of a received image with 20,000 elements. Discussion of the need for high-frequency transmission for real television.

1733 Farnsworth, Philo T., "Scanning With an Electric Pencil." Television News, 1:48-51, 74, March-April 1931. 4 photos, 2 diags. First full report on the Farnsworth system with details of "dissector" camera tubes and the "oscillite" receiver tube. Also, Dinsdale, Alfred, "Television by Cathode Ray." Wireless World, 28:286-288, March 1931.

1734 Halloran, Arthur H., "Scanning Without a Disc." Radio News, 12:998, 999, 1015, May 1931. 8 illus. Description of Farnsworth's system, with technical details of the dissector tube, circuit and waveforms.

1735 Zworykin, Vladimir K., "Television." J. F. I., 217:1-37, Jan. 1934. 31 illus. Outline of the all-electronic system developed by the author and the RCA Victor Company. Includes reproductions of received images and photos of an experimental television receiver. 9 footnotes.

1736 Zworykin, Vladimir K., "The Iconoscope--A Modern Version of the Electric Eye." Proc. IRE., 22:16-32, Jan. 1934. 14 illus. A basic paper by the inventor on the theory, construction, characteristics and operation of the new camera tube. Brief historical introduction.

1737 Bedford, Leslie H., and Owen S. Puckle, "A Velocity Modulation Television System." J. IEE., 75:63-82, discussion 83-92, July 1934. Illus. of experimental equipment and circuits. Discussion of theory, development, details and operation. Bibliography.

1738 Brolly, Archibald H., "Television by Electronic Methods." Elect. Eng., 53:1153-1160, Aug. 1934. 13 illus. Description of the Farnsworth system. 5 refs. in the text.

1739 Farnsworth, Philo T., "Television by Electron Image Scanning." J.F.I., 218:411-444, Oct. 1934. 21 diags., graphs and photos. Conversion of optical image to an electron image, magnetic focusing and deflection, image dissector tube, sensitivity, photomultiplier, oscillite receiver tube. Theory, constructions, operation. 3 refs.

1740 Hergenrother, Rudolph C., "The Farnsworth Electronic Television System." J. Television Soc., 1:384-387, 1934. 6 illus. 2 refs.

1741 Ardenne, Manfred von (Owen S. Puckle, trans.), Television Reception. London: Chapman and Hall, 1936. 121 pp., 43 plates and other illus. "Construction and operation of a cathode ray tube receiver for the reception of ultra-short wave television broadcasting." Technical treatment of cathode-ray tubes, description of receivers, tubes and circuits, with construction specifications and details.

1742 "Marconi-E.M.I. Television." J. Television Soc., 2:75-81, 1936. 5 illus. Description of the Marconi-E.M.I. equipment installed at Alexandra Palace, North London.

1743 McGee, James D., "Campbell Swinton and Television." Nature, 138:674-676, Oct. 17, 1936. A tribute to Swinton's proposals for electronic television. 3 extracts from published letters. 9 refs. from 1908. See 451.

1744 Murray, A.F., "The New Philco System of Televison."

Radio-Craft, 8:270, 315, Nov. 1936. 6 illus. General description with some technical details.

1745 [E. M. I.]. "Super-Emitron Camera." Wireless World, 41:497, 498, Nov. 18, 1937. Announcement of the camera tube developed by Electric and Musical Industries, Ltd.

1746 Blumlein, Alan D., and others, "The Marconi-E. M. I. Television System." J. IEE., 83:758-792, discussion, 793-801, Dec. 1938. Illus. Pt. 1, 758-766, Transmitted wave forms. Pt. 2, 767-782, C. O. Browne, Vision input equipment. Pt. 3, 782-792, N. E. Davis and E. Green, Radio transmitter.

1747 Iams, Harley, George A. Morton, and Vladimir K. Zworykin, "The Image Iconoscope." Proc. IRE., 27:541-547, Sept. 1939. 15 photos, graphs, diags. 11 footnote refs. from 1936. A paper presented at the IRE Convention, N. Y., June 17, 1938. Also, Electronics, 11:12, July 1938.

1748 McGee, James D., and Hans G. Lubszynski, "E. M. I. Cathode-Ray Television Transmission Tubes." J. IEE., 84:468-475, discussion and replies, pp. 475-482, April 1939. 8 illus., including 2 plates. A short history of the development of the Emitron and Super-Emitron camera tubes, with technical details, and a brief historical survey from 1908. 21 refs. from 1908 include a number of patents.

1749 Rose, Albert, and Harley A. Iams, "Television Pick-up Tubes Using Low-Velocity Electron-Beam Scanning." Proc. IRE., 27:547-555, Sept. 1939. 17 illus. Also, "The Orthicon, A Television Pick-up Tube." RCA Rev., 4:186-199, Oct. 1939. 11 photos and diags. 7 refs.

1750 Larson, C. C., and B. C. Gardner, "The Image Dissector." Electronics, 12:24-27, 50, Oct. 1939. 13 illus. Development, construction, operating principles and characteristics.

1751 [B. B. C.]. The London Television Station. London: British Broadcasting Corporation, 1939. 39 pp. Description of the facilities at Alexandra Palace, formally opened for regular transmission, Nov. 2, 1936.

1752 Zworykin, Vladimir K., and George A. Morton,

Television. The Electronics of Image Transmission. New York: John Wiley, 1940. 646 pp., 495 illus., 17 tables. Pt. 1, pp. 3-155, Fundamental physical principles, includes electron optics and vacuum practice. Pt. 2, pp. 159-262, Principles of television. Pt. 3, pp. 265-564, Component elements, deals with camera tubes, picture tubes, scanning, synchronization, amplifiers and receivers. Pt. 4, pp. 567-631, treats the RCA-NBC television project. 272 chap. refs. 2nd ed., 1954. 1037 pp., 697 illus., 17 tables. Arranged in 4 parts, as the first ed. Pt. 4, Color television, industrial television, and television systems. 465 chap. refs. A basic and important text with emphasis upon developments by RCA.

1753 Fink, Donald G., Principles of Television Engineering. New York: McGraw-Hill, 1940. 541 pp. 311 photos and diags. Technical and practical treatment of methods and equipment, image analysis, camera tubes, scanning beams, the video signal, video amplification, carrier transmission, image reproduction, broadcast practice, receivers. Some chap. bibliographies, numerous footnote refs. 2nd ed., 1952. 721 pp.

1754 Goldmark, Peter C., and others, "Color Television." Proc. IRE., 30:162-182, April 1942. Pt. 1, 33 diags., graphs and photos, table. 23 refs. to patents from 1904. 49 refs. to articles from 1910. 13 footnote refs. Also J. SMPE., 38:311-353, April 1942. Pt. 2, Proc. IRE., 31:465-478, Sept. 1943. 19 illus. 10 footnote refs.

1755 [Baird, John L.]. "New Baird Tube Gives Television in Color." Electronics, 17:190, 194, 198, Oct. 1944. 2 photos, diag. Description of the Telechrome color picture tube.

1756 Laden, Leon, "Television in Great Britain." Radio News, 33:32-34, 84, 86, Jan. 1945. 6 illus. Contemporary view, future plans, and Baird's color television equipment.

1757 Ehrlich, Robert W., "Color Television." Radio News, 34:32-34, 130, 132-138, July 1945. 14 illus., table of scanning frequencies. Details of the CBS system, and Baird's Telechrome tube. 8 refs.

1758 Rose, Albert, Paul K. Weimer, and Harold B. Law, "The Image Orthicon--A Sensitive Television Pickup

Tube." Proc. IRE., 34:424-432, July 1946. 10 photos and diags. 16 footnote refs.

1759 McGee, James D., "Electronic Generation of Television Signals." Electronics and Their Application in Industry and Research, Chap. 4, pp. 135-211. 28 photos and diags., 8 tables. Early history, pp. 135-149, has a reprint of Swinton's letter on all-electronic television (see 1728). The remainder deals with camera tubes; Farnsworth, EMI, RCA. 25 refs. from 1884, including patents. See 799.

1760 "R.C.A.'s Television." Fortune, 38:80-85, 194-204, Sept. 1948. 11 illus. History, production, investments, revenues, sales, business and market aspects.

1761 Parr, Geoffrey, "British Television." Discovery, 10(NS):212-218, July 1949. 13 illus. Some early history is followed by a brief description of the salient technical features of the EMI-BBC system. Reading list, 8 items from 1839.

1762 Preston, S. J., "The Birth of a High Definition Television System." J. Television Soc., 7:115-126, 1953. 8 illus. On the British development of the E.M.I. electronic television system, early Emitron camera tubes and BBC equipment. Decisions on an electronic system, technical details, camera tube development, circuit, choice of 405 lines, adoption of the E.M.I. system by the B.B.C., first outside broadcast, post-war advances, EMI-RCA relations and personalities (up to 1949).

1763 Weimer, Paul K., Stanley V. Forgue, and Robert R. Goodrich, "The Vidicon Photoconductive Camera Tube." Electronics, 23:70-73, May 1950. 5 photos and diags. 6 refs. from 1937.

1764 McGee, James D., "Distant Electric Vision." Proc. IRE., 35:596-608, June 1950. 18 illus. A brief historical introduction is followed by descriptions of the design and development of EMI camera tubes: Emitron, Super-Emitron, and the CPS Emitron.

1765 [I.R.E.]. "Color Television Issue." Proc. IRE., 39:1123-1360, Oct. 1951. 20 papers. Also, "Second Color Television Issue," 42:1-348, Jan. 1954. 49 papers. Includes a bibliography, pp. 344-348, 56 papers not included in either television issue.

1766 Bourton, K., Bibliography of Colour Television. London: Television Society, 1954, 1955, 1956. 34 pp. 650 items arranged chronologically, including associated topics, from 1860.

1767 Bello, Francis, "Color TV: Who'll Buy a Triumph?" Fortune, 52:136-139, 201, 202, 204, 206, Nov. 1955. 10 illus., including color displays of systems and tubes: shadow mask, post accelerator, Chromatron and Apple tubes. Brief history, description of several systems, company efforts, industry problems and prospects.

1768 Patchett, G. N., "Colour Television." J. Brit. IRE., 16:591-620, Nov. 1956. 42 photos, graphs, diags. A description of various systems, transmitting and receiving equipment, the N. T. S. C. system and British standards. Includes theory of colour mixing and colorimetry. Technical treatment with some historical matter. 155 refs. from 1940.

1769 Gabor, Dennis, "Colour TV." Endeavour, 21:25-34, Jan. 1962. 11 diags., color plate. On the development of various color systems and tubes with some historical background. Bibliography, 5 items from 1956.

1770 Mertz, Pierre, "Long-Haul Television Signal Transmission." J. SMPTE., 75:850-855, Sept. 1966. 5 diags. Brief review of wire services, technical aspects, and image transmission, with historical highlights. 25 refs. from 1927.

D. Facsimile

The early history of phototelegraphy includes the inventions of Alexander Bain (1843), Frederick Bakewell (1848), the first commercial system of Giovanni Caselli (1861), and other copying or writing telegraphs for reproducing script. The best historical account of phototelegraphy is a German text by A. Korn and B. Glatzel, Handbuch der Phototelegraphie und Teleautographie, published in Leipzig, 1911. It also treats early television; Chap. 6, pp. 417-484. Facsimile is covered in several items in Sect. 16A: 1638, 1653, 1654, 1658; and in Sect. 16B: 1672, 1693, 1699, 1709, 1711, 1712. See also 58, 64, 277, 423, 425, 1193.

1771 [Caselli, Giovanni]. "Autographic Telegraph." Elementary Treatise on Natural Philosophy, pp. 730-733. 4 illus. Description of Caselli's "Pantelegraph" and its operation. See 190.

1772 [Carey, George R.]. "Seeing by Electricity." Sci. Am., 42:355, June 5, 1880. 2 plates with 9 detailed views of the apparatus. Also, English Mechanic, 31:345, 346, June 18, 1880.

1773 [Senlecq, Constantin]. "The Telectroscope." English Mechanic, 32:534, 535, Feb. 11, 1881. 5 diags. Also, Sci. Am. Supp., 11:4382, 1881.

1774 Bidwell, Shelford, "Selenium and Its Applications to the Photophone and Telephotography." Proc. Roy. Inst., 9:524-535, 1881. Also, English Mechanic, 33:158, 159; 180, 181, April 22, 29, 1881. 4 diags. On selenium, its characteristics, construction of cells and demonstration of experimental apparatus. See also 744.

1775 [Knudson, Hans]. "Knudson's System of Wireless Transmission of Photographs." Electrician, 61:89, 90, May 1, 1908. 4 illus. On a demonstration in London, April 28.

1776 Armagnat, Henri, "Phototelegraphy." Smithsonian Institution, Annual Report for 1908, pp. 197-207 (Washington, 1909). Translated from the Revue Scientifique, April 18, 1908. 9 diags. Survey of current systems: Korn, Berjonneau, Carbonelle, Senlecq-Tival.

1777 Dubois, Louis, "Korn's Apparatus for Photographic Transmission." Electrician, 62:570-573; 644-647, Jan. 22, Feb. 5, 1909. 13 diags.

1778 Baker, T. Thorne, The Telegraphic Transmission of Photographs. London: Constable; New York: D. Van Nostrand, 1910. 141 pp., 66 illus. A survey of phototelegraphic systems, particularly Prof. Korn's and the author's apparatus. Chap. 7, pp. 127-141, covers the transmission of photographs and pictures by wireless telegraphy, with details of Knudson's work and the author's experiments. Numerous circuit diagrams and pictures of reproductions. Not documented except for some names and dates in the text. No index.

1779 Baker, T. Thorne, "The Telegraphy of Photographs,

Wireless and by Wire." Smithsonian Institution, Annual Report for 1910, pp. 257-274 (Washington, 1911). 12 illus., including 2 plates. Details of Korn's system and the author's wireless method, the Paris-London service inaugurated by the Daily Mirror, Nov. 1907, and the London-Manchester tests. See also, Nature, 82:309-311, Jan. 13; 84: 220-226, Aug. 18, 1910.

1780 Korn, Arthur, "The Transmission of Photographs and Drawings by Wireless Telegraphy." Wireless World, 1:353-357, Sept. 1913. 6 illus., including a circuit diag. Also p. 246, Oct. 1913.

1781 "Facsimile Telegraphy and Phototelegraphy." Sci. Am., 112:571-573, June 5, 1915. A summary of progress from 1843 with about 70 names, dates and brief descriptions. See 277.

1782 Arapu, R., "The Telephotographic Apparatus of G. Rignoux." Sci. Am. Supp., 79:331, May 22, 1915. "Experiments in sending visible forms by electricity." Diag. of system.

1783 Martin, Marcus J., "Radiophotography." Wireless World, 2:656-658; 727-731; 755-759; 3:57-60; 102-106; 162-165; 228-232, Jan.-July, 1915. Photos, diags. General discussion with mention of other systems, but mainly on the writer's work and machines. Also published as a book, The Wireless Transmission of Photographs. London: Wireless Press, 1916. 117 pp. 2nd ed., 1919. 145 pp., 77 illus.

1784 Gradenwitz, Alfred, "Radio-Telephotography." Radio News, 4:226, 227, Aug. 1922. 11 illus. On the Dieckmann radio system for the transmission of drawings and its use in aviation.

1785 Benington, Arthur, "Transmission of Photographs by Radio." Radio News, 4:230, 369-372, Aug. 1922. 2 illus. On the Korn coded system, with some historical background.

1786 Isakson, D. W., "Developments in Telephotography." Trans. AIEE., 41:794-801, Aug. 1922. 10 illus. A slight historical survey is followed by a description of the Leishman coded facsimile system.

1787 Davis, Watson, "Seeing by Radio." Popular Radio, 3:266-275, April 1923. 8 illus. Description of Jenkins' apparatus employing prismatic rings for transmitting pictures by radio.

1788 Winters, S.R., "The Transmission of Photographs by Radio." Radio News, 4:1772, 1773, April 1923. 5 illus. On the Jenkins system and demonstration in Washington, Oct. 3, 1922.

1789 Blake, George G., "Wireless Transmission of Photographs, or Telephotography." History of Radio Telegraphy and Telephony, pp. 220-222. Brief summary of events from the 1860s to 1923, with names, dates and refs. See 1193.

1790 [AT&T]. "Telephoning our Press Photographs." Sci. Am., 131:87, 139, Aug. 1924. 3 illus. Demonstration of the AT&T system.

1791 Herndon, Charles A., "1,000 Printed Words a Minute by Radio." Popular Radio, 7:11-15, Jan. 1925. 5 illus. On the Jenkins system employing prismatic disks and photographic recording film.

1792 [RCA]. "The Photoradiogram System in Trans-Atlantic Tests." Radio News, 6:1385, 1456, 1458, 1460, Feb. 1925. 5 illus. On the demonstration tests between New York and London, Nov. 30, 1924, with brief description of apparatus and operation.

1793 King, R.W., "Color Pictures by Radio." Popular Radio, 7:125-128, Feb. 1925. 3 illus. Bell System experiments. See next entry.

1794 Ives, Herbert E., and others, "The Transmission of Pictures Over Telephone Lines." B.S.T.J., 4:187-214, April 1925. 22 illus., including examples of reproductions. Frontispiece, example of 3-color photo reproduction. 6 footnote refs.

1795 Henry, Charles C., "A New Method of Transmitting Pictures by Wire or Radio." Radio Broadcast, 7:18-28, May 1925. 11 illus. On the Cooley system.

1796 Hansen, Edmund H., "The New Bartlane System of Radio Transmission of Pictures." Popular Radio,

7:(406), 407-411, May 1925. 4 illus. On the Bartholomew and McFarlane system.

1797 Jenkins, C. Francis, Vision by Radio. Radio Photographs. Radio Photograms. Washington: Jenkins Laboratories, 1925. 140 pp., port., 65 illus. Miscellaneous collection of pictures, with supporting text, mainly of the author's apparatus. Historical sketch of Jenkins' radio photography from 1894, pp. 118, 119. List of radio patents of interest, p. 132, 62 various items. No index.

1798 Ranger, Richard H., "Transmission and Reception of Photoradiograms." Proc. IRE., 14:161-180, April 1926. 19 photos, diags., and pictures of early reproductions. Brief history and details of the RCA equipment.

1799 Baker, T. Thorne, Wireless Pictures and Television. A Practical Description of the Telegraphy of Pictures, Photographs and Visual Images. London: Constable, 1926; New York: D. Van Nostrand, 1927. 188 pp., 99 illus. Much historical material by a pioneer in phototelegraphy. Almost all on facsimile; television, pp. 167-184. Over 50 footnotes.

1800 [Jenkins, C. Francis]. "Weather Maps by Radio." Sci. Am., 136:411, 427, June 1927. Two reproductions of weather maps by the Jenkins system.

1801 Baker, T. Thorne, "Phototelegraphy." Experimental Wireless, 4:229-238, 1927. On the work of Arthur Korn and his current system.

1802 Secor, H. Winfield, and J. H. Kraus, "Telegraphing Photos by Code." All About Television, Chap. 5, pp. 22-29, 108. 23 illus. General description of various systems: Belin, Andersen, Leishman, Korn, Bartlane, and earlier history. Chap. 6, pp. 30-59, 108, "Direct Wire or Radio Picture Transmission (Codeless)." 115 illus. As previous chapter, plus other systems: Dieckmann, Jenkins, Ferree, AT&T, RCA, Telefunken, and others. See 1693.

1803 [Alexanderson, Ernst F. W.]. "Home Radio Photography." Radio News, 9:1101-1103, 1163, April 1928. 9 illus. The Alexanderson (G. E.) system, demonstrated by NBC in New York. Also, Sci. Am., 138:338, April 1928. 5 illus.

1804 Haynes, F.H., "The Fultograph." Wireless World, 23:555-560, Oct. 27, 1928. Description of home radio facsimile receiver designed by Otto Fulton. Also, Radio News, 10:317, Oct. 1928.

1805 Ranger, Richard H., "Photoradio Developments." Proc. IRE., 17:966-984, June 1929. 13 photos, graphs, diags. Review of progress during 1926-1928 with mention of various systems; AT&T, Bartlane, Karolus-Telefunken, Marconi, Jenkins, Dieckmann, Cooley, RCA. No refs.

1806 Buehling, Norman D., "Telephotography. Transmission of Photographs by Wire." Radio-Craft, 3:724, 725, 747, June 1932. 3 photos, 4 diags. Discussion with technical details of the Bell Telephone Laboratories system. Complete schematic diagrams of transmitter and receiver.

1807 Cockaday, Laurence M., "If Not Television Why Not Facsimile." Radio News, 16:76, 77, 113, Aug. 1934. 4 illus. John V. L. Hogan's system and demonstration in New York.

1808 Callahan, John L., J.N. Whitaker, and Henry Shore, "Photoradio Apparatus and Operating Technique Improvements." Proc. IRE., 23:1441-1482, Dec. 1935. 26 photos, graphs and diags., 4 tables. Detailed description of RCA system, equipment, machines, circuits and operation. 9 footnote refs.

1809 Reynolds, F.W., "A New Telephotograph System." Elect. Eng., 55:996-1007, Sept. 1936. 13 photos and diags. Map of a leased wire network in U.S. Technical details of the Bell system with circuits. 2 pictures of Associated Press reproductions. 14 refs. from 1918.

1810 Young, Charles J., "Equipment and Methods Developed for Broadcast Facsimile Service." RCA Rev., 2:379-395, April 1938. 7 illus.

1811 Goldsmith, Alfred N., and others (eds.), Radio Facsimile. New York: RCA Institutes Technical Press, 1938. 353 pp. 20 papers by engineers of the RCA Laboratories. Pt. 1, pp. 1-111, Historical development, 6 papers. Pt. 2, pp. 112-197, Status in 1938, 3 papers. Pt. 3, pp. 198-304, Methods and equipment, 8 papers. Pt. 4, pp. 305-353, Radio facsimile broadcasting, 3 papers. Most of the papers originally appeared in Proc. IRE., and RCA Rev., several are entered separately in this section. A bibliography by J. L. Callahan is a most useful compilation (see next item).

1812 Callahan, John L., "A Narrative Bibliography of Radio Facsimile." Radio Facsimile, pp. 112-128 (see previous item). Brief narrative with lists of names and references in sections; historical, systems, apparatus, future trends. 261 refs. to books, articles and patents from 1895.

1813 Hogan, John V. L., "Facsimile and Its Future Uses." Ann. Am. Acad., 213:162-169, Jan. 1941. A general nontechnical survey.

1814 Lister, W. C., "The Development of Photo-Telegraphy." Electronic Engineering, 19:37-43, Feb. 1947. 13 illus. Brief review of some early historical apparatus and 20th-century systems. No refs.

1815 Lauden, Franklyn K., "Radio Facsimile May Print Newspapers of Tomorrow." Radio and Television News, 40:39, 148, Aug. 1948. 3 illus. Brief details of the Hogan system and demonstration by WQXR-FM, The New York Times station, Feb. 24, 1948.

1816 [RCA]. Ultrafax. New York: Radio Corporation of America, 1948. 32 pp., 25 illus. of equipment, circuits and reproductions of transmitted material. A publicity brochure with brief description of the system operation.

1817 McKenzie, A. A., "Ultrafax." Electronics, 22:77-79, Jan. 1949. 5 illus. New facsimile system employing cathode-ray tubes and film at transmitter and receiver demonstrated by RCA and Eastman Kodak in Washington, Oct. 21, 1948.

1818 Bond, Donald S., and Vernon J. Duke, "Ultrafax." RCA Rev., 10:99-115, March 1949. 6 photos and diags., table. Technical description of high-speed facsimile system employing electronic scanning and other television techniques. 8 footnote refs. Also, J. Brit. IRE., 9:146-156, April 1949.

1819 Jones, Charles R., Facsimile. New York: Murray Hill Books, 1949. 422 pp., 223 illus. A complete review of current systems with technical details and description of commercial equipment. Chap. 1, pp. 1-23, Facsimile in the past, is a historical survey from Alexander Bain up to 1930. 16 illus., 18 footnote refs. to British and U.S. patents.

1820 Cole, A. W., and J. A. Smale, "The Transmission of Pictures by Radio." Proc. IEE., 99(Pt. 3):325-335, discussion pp. 359-363, Nov. 1952. 9 illus. Review since 1842, particularly of radio methods from 1924, with details of British apparatus. 10 refs.

AUTHOR INDEX

Entries in Sect. 2E, Magazines and Journals (98-188) are excluded from this index. Unsigned articles, editorial notices, and other group publications are entered under the name of the publication, society or organization.

SUBJECT INDEX

This listing includes the works as well as the names of men and companies mentioned in the entries. See also the related entries in the author index. For subject groups see Contents. For individual topic, see under main subject heading, e.g., Electron tube, Radio, Television. Some items are indexed under two or more topics.